Principles of Blockchain Systems

Synthesis Lectures on Computer Science

Principles of Blockchain Systems
Antonio Fernández Anta, Chryssis Georgiou, Maurice Herlihy, Maria Potop-Butucaru
2021

Automated Verification of Concurrent Search Structures
Siddharth Krishna, Nisarg Patel, Dennis Shasha, and Thomas Wies
2021

Creating Autonomous Vehicle Systems, Second Edition
Shaoshan Liu, Liyun Li, Jie Tang, Shuang Wu, and Jean-Luc Gaudiot
2020

Blockchain Platforms: A Look at the Underbelly of Distributed Platforms
Stijn Van Hijfte
2020

Analytical Performance Modeling for Computer Systems, Third Edition
Y.C. Tay
2018

Creating Autonomous Vehicle Systems
Shaoshan Liu, Liyun Li, Jie Tang, Shuang Wu, and Jean-Luc Gaudiot
2017

Introduction to Logic, Third Edition
Michael Genesereth and Eric Kao
2016

Analytical Performance Modeling for Computer Systems, Second Edition
Y.C. Tay
2013

Introduction to Logic, Second Edition
Michael Genesereth and Eric Kao
2013

Introduction to Logic
Michael Genesereth and Eric Kao
2013

Principles of Blockchain Systems

Antonio Fernández Anta, Chryssis Georgiou, Maurice Herlihy, Maria Potop-Butucaru

ISBN: 978-3-031-00679-1   paperback
ISBN: 978-3-031-01807-7   ebook
ISBN: 978-3-031-00075-1   hard
DOI: 10.1007/978-3-031-01807-7

A Publication in the Springer series
SYNTHESIS LECTURES ON COMPUTER SCIENCE
Series ISSN: 1932-1228 print   1932-1686 ebook

Lecture #14

First Edition
10 9 8 7 6 5 4 3 2 1

Principles of Blockchain Systems

Antonio Fernández Anta
IMDEA Networks Institute, Madrid, Spain

Chryssis Georgiou
University of Cyprus, Nicosia, Cyprus

Maurice Herlihy
Brown University, Providence, RI, USA

Maria Potop-Butucaru
Sorbonne University, Paris, France

SYNTHESIS LECTURES ON COMPUTER SCIENCE #14

ABSTRACT

This book is the first to present the state of the art and provide technical focus on the latest advances in the foundations of blockchain systems. It is a collaborative work between specialists in cryptography, distributed systems, formal languages, and economics, and addresses hot topics in blockchains from a theoretical perspective: cryptographic primitives, consensus, formalization of blockchain properties, game theory applied to blockchains, and economical issues.

This book reflects the expertise of the various authors, and is intended to benefit researchers, students, and engineers who seek an understanding of the theoretical foundations of blockchains.

KEYWORDS

Blockchain, Cryptography, Consensus, Distributed Computing, Distributed Ledger Technology, Foundations, Formal Methods, Game Theory, ICOs, Interoperability, Mining Pools, Principles, Smart Contracts, Tokens

Contents

1. **Cryptographic Tools for Blockchains** 1

Björn Tackmann, DFINITY Foundation, Zürich, Switzerland
Ivan Visconti, University of Salerno, Salerno, Italy

2. **A Consensus Taxonomy in the Blockchain Era** 27

Juan A. Garay, Texas A&M University, College Station, TX, USA
Aggelos Kiayias, University of Edinburgh, Edinburgh, UK

Ilya Sergey, Yale-NUS College and National University of Singapore, Singapore

Preface

The blockchain phenomenon resembles a 19th century gold rush. Blockchain technologies are sold, and perhaps sometimes oversold, as a way to decentralize activities such as finance and administration, activities that have been centralized for centuries.

A blockchain is a distributed ledger. Like classical ledgers, a blockchain is a transparent, tamper-proof sequence of records. Unlike classical ledgers, a blockchain operates in an untrusted, distributed environment. Specifically, a blockchain system maintains a continuously growing sequence of blocks, each of which records one or more transactions verified by the members of the system. Cryptographic techniques ensure that blocks are tamper-proof, and the blocks' contents and their order are determined by a distributed consensus algorithm run by the system participants. The consensus algorithm allows the participants to co-operate without necessarily trusting one another. Blockchains must operate in a complex environment subject to faults, including malicious and even irrational behavior.

The scientific study of blockchain systems raises issues in several disciplines ranging from technical aspects of distributed systems and programming to cryptography and privacy, game theory, and economic analysis, and even to legal and regulatory questions.

This book addresses the principles of blockchain systems from the perspective of computer science and economics. We bring together researchers from distributed computing, cryptography, game theory, programming and formal methods, and economics. This material is intended to be helpful for both practitioners and researchers, as well as for students seeking a path into this area. The range of potential applications is vast: digital decentralized finance (DeFi), healthcare, insurance, supply chain management, and so on.

Distributed ledger technology is a tree whose roots encompass cryptography, distributed systems and programming, game theory, and economics. To reflect this structure, this book has eight chapters, each dedicated to a specific root. Chapter 1 introduces the cryptographic tools used in blockchains, Chapter 2 is dedicated to a taxonomy of consensus algorithms, Chapter 3 discusses smart contracts and the link with programming and formal methods, Chapter 4 proposes new advances in the formalization of blockchain properties, Chapter 5 addresses adversarial interoperable cross-chain commerce, Chapter 6 surveys strategic interactions in blockchain protocols from the game-theoretic perspective, Chapter 7 focuses on game theory applied to Bitcoin mining pools, and finally, Chapter 8 overviews the economic literature on tokens and ICOs. In the following, we provide a flavor of each one of the eight chapters of the book.

Chapter 1: Cryptographic Tools for Blockchains

Chapter 1, authored by Björn Tackmann and Ivan Visconti, presents some of the most relevant cryptographic tools used to build blockchains and their applications. The chapter describes tools used in existing blockchains, as well as tools likely to be useful in the future.

Chapter 2: A Consensus Taxonomy in the Blockchain Era

Consensus is one of the most fundamental problems in distributed computing. In particular, it enables secure broadcast, which lies at the heart of many cryptographic protocols. Consensus has a long and rich history. With the advent of blockchain systems, consensus has experienced a new surge of interest and an explosion of novel applications.

Consensus takes many different forms, varying with application requirements, computational assumptions, and network models. In Chapter 2, Juan Garay and Aggelos Kiayias present a systematization of knowledge, encompassing the landscape of consensus research in the Byzantine failure model, starting with the original formulation in the early 1980s, extending to today's blockchain-based consensus protocols. This chapter is a roadmap for studying the consensus problem under its many guises, and its emerging blockchain-era applications.

Chapter 3: The Next 700 Smart Contract Languages

Smart contracts are special-purpose programs whose executions are replicated by means of a distributed consensus protocol. The parties participating in the consensus protocol can, thus, define custom logic for transactions by deploying smart contracts that implement arbitrary applications to be executed in a decentralized way. Smart contracts have proven useful for distributed finance, including digital accounting, voting, and schemas for distribution of assets.

Smart contracts have attracted interest from the research community due to their correctness-critical nature, and also because of a number of costly accidents that involved deployed faulty contract implementations. In retrospect, many of those issues could have been avoided with a more careful choice of linguistic abstractions—that is, with a programming language design tailored for the domain.

In Chapter 3, Ilya Sergey provides a high-level overview of the design choices in programming languages for smart contracts dictated by their essential aspects: (i) atomicity, (ii) communication, (iii) management of digital assets, and (iv) resource accounting. He argues that finding a balancing act for expressing those concepts poses a fundamental challenge in the programming language design for blockchain-based decentralized applications.

Chapter 4. Formalization of Blockchain Properties

As token economy and cryptocurrencies thrive, blockchains have become popular in several scientific and application domains. Although many of the functions provided by blockchains are widely appreciated, many domain experts rely on informal characterizations of their systems, often not clearly differentiating between a cryptocurrency, the ledger that supports it, or the services it provides. In many cases, the code itself is the only specification provided: that is, "the code is the spec."

In Chapter 4, Emmanuelle Anceaume, Antonio Fernández Anta, Chryssis Georgiou, Nicolas Nicolaou, and Maria Potop-Butucaru attempt to decipher the fundamentals behind the distributed aspects of blockchains. They provide *formal specifications* and *specific properties* of the underlying data structures. In particular, they present recent attempts to formalize the properties of distributed ledgers and blockchains from a distributed computing point of view. The authors first examine permissioned distributed ledgers as an abstract data type, and they formally present the operations and value domains it supports. Treating the distributed ledger as a shared object, they define what it means for the ledger to be consistent. They introduce the concept of the blockchain abstract data type, which provides a lower-level abstraction of distributed ledgers, suitable for both permissioned and permissionless systems.

Chapter 5: Adversarial Cross-Chain Commerce

How can autonomous, mutually distrusting parties cooperate safely and effectively? Classical distributed systems have proposed ways to combine multiple steps into a single atomic action, to recover from failures, and to synchronize concurrent access to data. Nevertheless, each of these issues requires rethinking when participants are autonomous and potentially adversarial. A *cross-chain deal* is a new way to structure complex distributed computations that manage assets in an adversarial setting. Deals are inspired by classical atomic transactions, but are necessarily different, in important ways, to accommodate the decentralized and untrusting nature of the exchange. In Chapter 5, Maurice Herlihy, Barbara Liskov, and Liuba Shrira explore how to think about multi-party cross-chain deals that span multiple blockchains.

Chapter 6: Strategic Interactions in Blockchain: A Survey of Game-Theoretic Approaches

In Chapter 6, Bruno Biais, Christophe Bisière, Matthieu Bouvard, and Catherine Casamatta survey *game-theoretical* studies of interactions in the blockchain. Most of the contributions they consider are devoted to proof of work (PoW). They first review analyses of mining strategies, studying whether miners will follow the longest chain rule or fork, and whether undesirable strategies such as selfish mining or double spending can occur in equilibrium. Second, they present analyses of the supply of mining services, such as computing capacity choices and the organization of mining pools, studying whether equilibrium outcomes are efficient. Third, they discuss analyses of the determinants of transaction fees, and their consequences on network congestion and the allocation of priority among users. Finally, they point to the nascent literature on consensus in alternative protocols such as proof of stake.

Chapter 7: Bankruptcy Solutions as Reward Functions in Mining Pools

In the Bitcoin system, mining is the procedure through which miners can gain money on a regular basis by solving mathematical crypto-puzzles that validate a Bitcoin transaction block. This situation can be described as a *cooperative game* where each mining pool receives a reward that has to be split among the pool's participants. The main challenge here is *how to divide it*. When put in terms of game theory, this challenge becomes *how to redistribute the amount gained by a coalition between its components.*

There exist, in the literature and in practice, several reward functions that allocate Bitcoins in pools. In Chapter 7, Marianna Belotti and Stefano Moretti extend a reward function proposed in the literature by combining it with two well-known solutions from bankruptcy situations: the *constrained equal awards* (CEA) rule and the *constrained equal losses* (CEL) rule. Using a property-driven approach, they argue that an *incentive-compatible* reward function from the literature guarantees a good intra-pool behavior, whereas the CEL rule is more adapted to prevent malicious inter-pool behaviors.

Chapter 8: Tokens and ICOs: A Review of the Economic Literature

In Chapter 8, Andrea Canidio, Vincent Danos, Stefania Marcassa, and Julien Prat describe how *initial coin offerings* (ICOs) are conducted and provide an overview of data on their market history. They discuss the corporate finance issues raised by ICOs and how they should be designed in order to align the interests of sellers and buyers of digital coins. Finally, they present a few models of ICO pricing and conclude by discussing the numerous challenges that still have to be tackled.

A final note

The topic of distributed ledgers and blockchains is broad and encompasses multiple disciplines, so it is impossible to cover all aspects. Here we focus on the topics we consider to be the most foundational. We intend to update future editions of this book with new material, and perhaps new topics, as the field progresses. We hope you will find this book useful and enlightening. Enjoy!

Antonio Fernández Anta, Chryssis Georgiou, Maurice Herlihy, Maria Potop-Butucaru
August 2021

Acknowledgments

First, we would like to thank all the contributing authors. Without their work, this book would not have been possible. Namely, our thanks to Emmanuelle Anceaume, Marianna Belotti, Bruno Biais, Christophe Bisière, Matthieu Bouvard, Andrea Canidio, Catherine Casamatta, Vincent Danos, Juan Garay, Aggelos Kiayias, Barbara Liskov, Stefania Marcassa, Stefano Moretti, Nicolas Nicolaou, Julien Prat, Ilya Sergey, Liuba Shrira, Björn Tackmann, and Ivan Visconti. We also thank them for cross-reading one another's chapters and providing fruitful feedback that has helped improve each chapter, and thus the book as a whole. Last but certainly not least, we would like to thank our managing publishing editor, Diane Cerra, who was highly supportive from the very beginning. Thank you, Diane, for your persistence and for keeping on "pushing" us to complete the work in a (relatively) timely manner!

CHAPTER 1

Cryptographic Tools for Blockchains

Björn Tackmann, *DFINITY Foundation, Zürich, Switzerland*

Ivan Visconti, *University of Salerno, Salerno, Italy*

1.1 INTRODUCTION

Cryptography is a vital building block in every blockchain system: The individual blocks are chained by means of a hash function. All consensus protocols depend on some types of cryptographic schemes, such as, again, hash functions in proof-of-work consensus. Transactions need to be authorized by the appropriate stakeholders; this is usually achieved through digital signature schemes. Newer blockchain systems also make use of more complex cryptographic building blocks such as verifiable random functions in several proof-of-stake consensus protocols, or non-interactive zero-knowledge proofs for private but publicly verifiable transactions. The purpose of this chapter is to describe the types of cryptographic schemes relevant to state-of-the-art blockchain systems and their use within those systems. For an excellent introduction to the field of modern cryptography, we refer the reader to the Katz and Lindell [2014] textbook.

We follow the widely used practice of using asymptotics for algorithm complexity and security definitions. All algorithms, including those comprising the cryptographic schemes as well as those of the adversary attacking them, are described as probabilistic Turing machines that get as an additional input the so-called *security parameter* $\kappa \in \mathbb{N}$, and are required to run in polynomial time in κ as well as possibly the length of other inputs they obtain. Such Turing machines are referred to as *efficient* or *probabilistic polynomial time* (PPT) Turing machines. A function $f : \mathbb{N} \to [0, 1]$, $k \mapsto f(k)$ is called *negligible in k* if it vanishes faster than the inverse of any polynomial, that is, $\forall c \in \mathbb{N}\ \exists k_0 \in \mathbb{N}\ \forall k \geq k_0 : f(k) < k^{-c}$. Generally, a cryptographic scheme will be considered secure if no PPT adversary can break it with non-negligible success probability.

Most cryptographic schemes are only secure under the assumption that some computational problem is difficult to solve. Schemes based on RSA [Rivest, Shamir, and Adleman 1978] require some variant of the assumption that for two large primes $p, q \in \mathbb{N}$, given only their product $N = pq \in \mathbb{N}$, finding p and q is hard. Schemes based on the ideas of Diffie and Hellman [1976] require some variant of the discrete-logarithm assumption; namely, that for a given group $\mathcal{G} = \langle g \rangle$

with generator g, for a uniformly random $x \in \{1, \ldots, |\mathcal{G}|\}$ and given only g and $g^x \in \mathcal{G}$, it is hard to find x. Discrete logarithm-based schemes have long used (a large subgroup of) the multiplicative group $\mathcal{G} \subset \mathbb{Z}_p^\times$ for a specific type of prime $p \in \mathbb{N}$, but recently elliptic-curve based schemes have become more popular, especially due to the smaller key sizes they allow. Another attractive property of certain elliptic curves is that they may admit bilinear pairings; namely, for three elliptic curve groups $\mathcal{G}_1, \mathcal{G}_2, \mathcal{G}_T$ there is an efficiently computable map $e : \mathcal{G}_1 \times \mathcal{G}_2 \to \mathcal{G}_T$ that is linear in both arguments. Starting with the work of Boneh and Franklin [2001], pairings have been used to describe efficient schemes for tasks that are more difficult to achieve by means of the other widely used types of assumptions. In all above cases, the asymptotic versions of the computational assumptions refer to a sequence of the described computational problems, one for each value of the security parameter.

Further notation

The set of all bit strings is denoted by $\{0, 1\}^*$. The concatenation of two bit strings $x, y \in \{0, 1\}^*$ is denoted by $x \mid y \in \{0, 1\}^*$.

1.2   HASH FUNCTIONS AND APPLICATIONS

Hash functions are a core cryptographic primitive in blockchain systems, since they link the individual blocks by including in each block a hash of its parent (the previous) block. A secure hash function guarantees that it is infeasible to find a different parent block that hashes to the same value; this is sufficient to create a chain that unequivocally links each block in the chain from the genesis block to the current tip. Hash functions also play a crucial role in many more complex cryptographic schemes and protocols that are discussed later in this chapter.

1.2.1   HASH FUNCTIONS

At an intuitive level, a hash function is a function that maps an input value of arbitrary length to a short output value, such that it is computationally difficult to find a collision, which is defined as a pair of inputs that hash to the same output. The difficulty of finding a collision for a fixed function $H : \{0, 1\}^* \to \{0, 1\}^\ell$ for output length $\ell \in \mathbb{N}$ is not easy to define: By a simple counting argument there exist $x, y \in \{0, 1\}^*$ with $H(x) = H(y)$, and so there exists an algorithm that produces a collision, namely the one that outputs x and y. Therefore, the widely used model in cryptography is that one assumes a hash function to be chosen at random from a given family of functions. Alternatively, one can define a hash function as a function that additionally uses a key.

A bit more concretely, a *(keyed) hash function family with output length* $\ell : \mathbb{N} \to \mathbb{N}$ is a sequence $\{H_\kappa\}_{\kappa \in \mathbb{N}}$ such that, for each security parameter $\kappa \in \mathbb{N}$, H_κ is an efficiently computable function $H_\kappa : \{0, 1\}^\kappa \times \{0, 1\}^* \to \{0, 1\}^{\ell(\kappa)}$. Collision resistance of such a hash function family is then defined using a simple random experiment. Initially, a hash-function key $K \in \{0, 1\}^\kappa$ is chosen at random.

If there exists no efficient algorithm $\mathcal{A}$ that, on input K, outputs a collision with probability that is non-negligible in κ, then the hash function family is called *collision resistant*.

The cryptographic definition of keyed hash functions is a bit at odds with hash functions used in practice. In fact, the most widely used hash functions in practice are nowadays SHA-2 [NIST 2015a] and SHA-3 [NIST 2015b], but other candidates from the SHA-3 competition are also used. Practical hash functions are fixed functions; they only exist for a few output lengths and it is unclear whether a random key was chosen in their design. For notational simplicity and consistency with practice, we will generally denote the application of a hash function to a bit string x as $H(x)$, keeping in mind that making the security statements formal often requires the keyed version.

The random oracle model

It is convenient in the analysis of cryptographic mechanisms to model a hash function with output length ℓ as a perfectly random function $\{0, 1\}^* \to \{0, 1\}^\ell$. This idealization is referred to as the *random oracle model* [Bellare and Rogaway 1993]. While there are constructions of protocols that are secure in the random-oracle model but insecure when the random oracle is instantiated with any concrete hash function [Canetti et al. 2004], the random-oracle model still appears to be a useful heuristic for analyzing the security of protocols.

Example 1.1 (The blockchain)   The most fundamental application of hash functions in blockchain systems is in chaining the blocks, as depicted in Figure 1.1. The blockchain consists of a sequence of blocks $B_0, B_1, \ldots$ that contain transactions. Each block B_i also contains a hash $H(B_{i-1})$ of its parent block B_{i-1}, the only exception being the genesis block B_0. The effect of chaining the blocks is that every participant that has a correct copy of the latest block, such as B_{i+2} in Figure 1.1, can validate the correctness of the entire chain $B_0, \ldots, B_{i+2}$. An incorrect block is either detected—or would immediately yield a collision in the hash function.

Example 1.2 (Addresses in cryptocurrencies)   Most of the widely used cryptocurrencies use the concept of an *address* to identify the source and the destination of a token (or coin) transfer. Each token is controlled by a private/public key pair, as described in Section 1.3. Starting with Bitcoin, most cryptocurrency platforms define the address based on the hash of the public key, in a specific

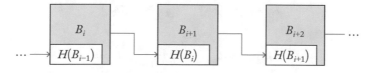

FIGURE 1.1: Chain of blocks: Each block B_i contains a hash of its parent block B_{i-1}.

printable encoding. This approach is sound because the collision resistance of the hash function guarantees that it is hard to find a second public key that hashes to the same address.

1.2.2   MERKLE TREES

A *Merkle tree* or *hash tree* is a data structure that allows authentication of a list of values through a single hash, while still allowing efficient verification of individual elements of the list [Merkle 1980]. Consider bit strings $x_1, \ldots, x_N \in \{0, 1\}^*$, where for simplicity we assume that $N = 2^n$ for some $n \in \mathbb{N}$, and let H be a hash function with output length ℓ. A *binary hash tree* (or Merkle tree) is then built as follows: The leaves of the tree, which we also refer to as level-n nodes, are computed as $h_{n,i} = H(x_i)$ for $i \in \{0, \ldots, N\}$. Given all nodes at a level i, one proceeds to compute the nodes at level $i - 1$ as $h_{i-1,j} = H(h_{i,2j-1}|h_{i,2j})$, for $j = 1, \ldots, 2^{i-1}$ until one ends up at a single root $h_{0,1} = H(h_{1,1}|h_{1,2})$. A binary hash tree of depth 2 is depicted in Figure 1.2.

If the hash function H is collision resistant, then the root of the Merkle tree indeed authenticates all leaves. To see this, assume that two different lists of leaves lead to the same root, that is, the lists differ in some position k. Along the path from the *different* leaves in position k to the *same* root is at least one level where *different* inputs to H lead to the *same* output; this position contains a collision for H.

A key property of the Merkle tree is that one can prove membership for some x_k in length ℓn. Define the helper function $f : \mathbb{N} \to \mathbb{N}$ that maps $i \mapsto i + 1$ if i is odd and $i \mapsto i - 1$ if i is even. The membership proof for x_k then consists of the elements $h_{n, f(k)}, h_{n-1, f(\lceil k/2 \rceil)}, h_{n-2, f(\lceil k/4 \rceil)}, \ldots,$ $h_{1, \lceil 2k/N \rceil}$. The verification proceeds by re-computing the tree along the path from $h_k = H(x_k)$ to the root: the elements in the proof allow application of the recursive formula for $h_{i-1,j}$ provided above. As an example, consider again the tree from Figure 1.2 and suppose we want to prove that x_2 is in position 2 of the Merkle tree with root $h_{0,1}$. The proof consists of elements $h_{2,1}$ and $h_{1,2}$; it is easy to verify that x_2, $h_{2,1}$, and $h_{1,2}$ suffice to re-compute $h_{0,1} = H(H(h_{2,1} \mid H(x_2)) \mid h_{1,2})$.

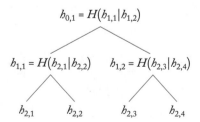

FIGURE 1.2: A binary Merkle tree of depth 2.

Example 1.3 (Use of Merkle trees in Zerocash)   The root of a Merkle tree succinctly authenticates the list of all leaves. This is used, for instance, in the Zerocash protocol [Ben-Sasson et al. 2014a], in which each leaf corresponds to a coin. A transaction spending a coin contains a proof that the coin is indeed authenticated by the Merkle tree root. To make the scheme privacy-preserving, Zerocash uses commitments and non-interactive zero-knowledge proofs, which we explain in Sections 1.5 and 1.6.

1.3   DIGITAL SIGNATURES AND VARIANTS

A digital signature scheme allows one party, the *signer S*, to produce for a given message $m \in \{0, 1\}^*$ a bit string $s \in \{0, 1\}^\kappa$ of length κ, referred to as the *signature*, that allows the *verifier V* to check whether m was indeed signed by S. To this end, S owns a pair of keys, consisting of a *signing* key sk and a *verification* key vk. While sk must be secret and is used in the generation of the signature, vk is public and must be known by anyone who wishes to verify the authenticity of the signature. Digital signatures were initially proposed by Diffie and Hellman [1976]; early instantiations have been provided by Rivest, Shamir, and Adleman [1978], Lamport [1979], and Merkle [1979].

1.3.1   PLAIN DIGITAL SIGNATURES

A digital signature scheme formally consists of three algorithms for key generation, signature generation, and signature verification, respectively. The key-generation algorithm, on input of the security parameter κ, outputs a pair (sk, vk) of *signing key sk* and *verification key vk*. The (possibly probabilistic) signing algorithm takes as input the signing key sk and a message $m \in \{0, 1\}^*$, and outputs a signature s. The (deterministic) verification algorithm takes as input the verification key vk, a message m, and the purported signature s, and outputs a Boolean value that signifies whether the signature is valid. Correctness of a signature scheme means that honestly generated signatures always verify correctly.

The first formal security definition for digital signatures was given by Goldwasser et al. [1988]. Their definition, called *existential unforgeability under chosen-message attack (EUF-CMA)*, is described as follows. Let A be an algorithm that obtains the verification key vk of a digital signature scheme and has access to an oracle that provides, on input of a message $m \in \{0, 1\}^*$, a corresponding signature $s \in \{0, 1\}^\kappa$. A *forgery* is a pair of message $\tilde{m}$ and signature $\tilde{s}$ such that $\tilde{s}$ is a valid signature for $\tilde{m}$ but $\tilde{m}$ has not been queried to the oracle. A digital signature scheme is said to achieve EUF-CMA if, for any efficient algorithm A, the probability of providing a forgery is negligible.

The signature scheme DS is said to be *strongly unforgeable under chosen-message attack (SUF-CMA)* if pairs $(\tilde{m}, \tilde{s})$ are also considered forgeries if $\tilde{m}$ has been queried to the oracle, as long as $\tilde{s}$ has not been returned as a signature for $\tilde{m}$ by the oracle.

Signature schemes used in practice

The signature schemes still most widely used overall are based on RSA, with its currently preferred variant PSS based on the work of Bellare and Rogaway [1996], whose security is based on the difficulty of factoring. The most widely used signature scheme whose security is based on the difficulty of computing discrete logarithms is DSA (Digital Signature Algorithm), which was standardized by NIST [2013]. In the context of blockchains, its variant ECDSA, which is based on elliptic curves, is more prominent due to its smaller key and signature sizes. Another variant based on discrete logarithms is Schnorr signatures [Schnorr 1989], especially due to their use in EdDSA (Edwards-curve Digital Signature Algorithm [Bernstein et al. 2011]). A different scheme that is particularly interesting due to its variants with extended functionality was introduced by Boneh et al. [2001], and is often referred to as *BLS* (for Boneh-Lynn-Shacham).

We provide a short overview of the BLS scheme here, as it serves as a basis for further, more complex schemes that we describe in the following sections. The scheme can be defined in any so-called *Gap Diffie-Hellman group*, in which computing g^{xy} from g^x and g^y is hard, but, given g^x, g^y, and $\tilde{g}$, deciding whether $\tilde{g} = g^{xy}$—whether $(g^x, g^y, \tilde{g})$ is a Diffie-Hellman triple—is easy. We describe the scheme in the bilinear pairing setting introduced in Section 1.1, as this will be helpful in the later sections. In that setting, the corresponding *co-Gap Diffie-Hellman assumption* states that for $\mathcal{G}_1 = \langle g_1 \rangle$ and $\mathcal{G}_2 = \langle g_2 \rangle$, computing h^a from $h \in \mathcal{G}_1$ and $g_2, g_2^a \in \mathcal{G}_2$ is hard, while deciding for given (g_2, g_2^a, h^b) whether $h^a = h^b$—whether (g_2, g_2^a, h^b) is a co-Diffie-Hellman triple—is easy. This test can indeed be checked easily by computing whether $e(h, g_2^a)$, which by bilinearity is the same as $e(h^a, g_2)$, equals $e(h^b, g_2)$.

In more detail, the BLS scheme assumes a pair $(\mathcal{G}_1, \mathcal{G}_2)$ of co-Gap Diffie-Hellman groups of order q and a hash function $H : \{0, 1\}^* \rightarrow \mathcal{G}_1$. The signature key is a uniformly random $x \in \mathbb{Z}_q$ and the verification key is $g_2^x \in \mathcal{G}_2$. A message m is signed with signature key x by computing $s \leftarrow H(m)^x$. Note that $(g_2^x, H(m), s)$ form a co-Diffie-Hellman triple. The bilinear pairing allows one to check the signature s given only $H(m)$ and g^x, but the co-Gap Diffie-Hellman assumption implies that computing it based on the same values is hard.

Example 1.4 (The use of signature schemes in transactions)   Digital signatures are instrumental for authorizing transactions. We describe a simplified variant of Bitcoin transactions to exemplify the use of signatures. Bitcoin is based on a model called *unspent transaction output (UTXO)*, where each transaction is thought of as having *inputs* that are spent and *outputs* that can be spent in a subsequent transaction. The inputs are outputs, then, of previous transactions and they must be unspent, hence the name. We depict the data structure of a basic (single-input single-output)

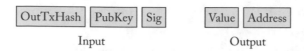

Input   Output

FIGURE 1.3: Data structure of a basic Bitcoin transaction (simplified). The input references the output of a previous transaction via the transactions hash (OutTxHash). It contains the signature public key (PubKey) and signature (Sig) to allow for verification. The output consists of the value (Value, in Bitcoin) it contains and the address (Address) of the public key that is allowed to spend it.

transaction in Figure 1.3. Bitcoin also supports multi-input multi-output transactions that consist of lists of inputs and outputs.

In the simplest case, each *output* is essentially a value in Bitcoin together with a Bitcoin *address*, that is, a hash of a verification key of a signature scheme. The meaning is that whoever controls the signature key associated with that address will be able to spend the denoted amount of Bitcoins in a future transaction. Each *input*, in the basic case, consists of a reference to a previous transaction output, a verification key, and an ECDSA signature. The verification key must hash to the address stated in the referenced output. The signature must be a valid signature for the remaining parts of the present transaction. That is, to create a transaction, one first generates the data structure without the signature, signs it, and then includes the signature before sending the transaction to the blockchain. Bitcoin transactions are considerably more flexible than that; outputs can specify scripts that describe the condition under which the output can be spent—generalizing the signature mechanism described here.

Many blockchain systems, including Bitcoin and Ethereum, use ECDSA as their signature algorithm. Bitcoin has chosen to use ECDSA with a different elliptic curve (often referred to as *secp256k1*) than that in the ECDSA standard (often referred to as *secp256r1*), and other blockchain platforms such as Ethereum have followed Bitcoin.

1.3.2 AGGREGATE SIGNATURES

An *aggregate signature scheme* enables a group of signers to combine their signatures on distinct messages into a compact representation. The first construction of aggregate signatures was given by Boneh et al. [2003] (BGLS, for Boneh-Gentry-Lynn-Shacham) and is based on pairings, which are based on the BLS scheme described in Section 1.3.1. BGLS signatures operate in the same bilinear pairing setting as described above, namely with groups $\mathcal{G}_1 = \langle g_1 \rangle$ and $\mathcal{G}_2 = \langle g_2 \rangle$ of order q, a target group $\mathcal{G}_T$, and a mapping $e : \mathcal{G}_1 \times \mathcal{G}_2 \to \mathcal{G}_T$. The hash function H maps into the group $\mathcal{G}_1$. Each user i has a signing key $sk_i = x_i \in \mathbb{Z}_q$ and a verification key $vk_i = g_2^{x_i}$. Individual signatures are computed as in BLS: a message $m \in \{0, 1\}^*$ is signed as $s_i \leftarrow H(m_i)^{x_i} \in \mathcal{G}_1$. Verification is computed using e, namely as $e(s_i, g_2) = e(H(m_i), vk_i)$.

The aggregation of signatures is surprisingly simple. For signatures $s_1, \ldots, s_n$ of n users on distinct messages $m_1, \ldots, m_n$, the aggregate signature is computed as $s \leftarrow \prod_{i=1}^{n} s_i \in \mathcal{G}_1$. Verification starts by first checking that all messages $m_1, \ldots, m_n$ are distinct, and then checking whether $e(s, g_2) = \prod_{i=1}^{n} e(H(m_i), vk_i)$ holds. That last equation makes sense since

$$e(s, g_2) = e\left(\prod_{i=1}^{n} H(m_i)^{x_i}, g_2\right) = \prod_{i=1}^{n} e(H(m_i)^{x_i}, g_2) = \prod_{i=1}^{n} e(H(m_i), g_2^{x_i}).$$

The scheme is restricted in the sense that the messages contributed by different parties must differ, which means that it is not a multi-signature scheme. The reason is that for equal messages a signer could choose their verification key adaptively and based on the verification key of another user, similarly to the attacks on early multi-signature schemes [Horster et al. 1995]. This attack is not possible if the group elements $H(m_i)$ used by the different signers differ, and indeed the scheme is provably secure under the co-Gap Diffie-Hellman assumption.

Bellare et al. [2007] later proved that the condition of distinctness can be dropped if instead of $H(m)^x$, each user computes their signature as $H(vk \mid m)^x$, which was conjectured by Boneh et al. [2003].

Sequential aggregate signatures

Introduced by Lysyanskaya et al. [2004], a sequential aggregate signature scheme achieves the same security as a (general) aggregate signature scheme, but requires that the signatures be aggregated in a certain predetermined order. This restriction allows for more efficient implementations; in particular, they show a scheme based on certified trapdoor permutations that is secure in the random-oracle model. Certified trapdoor permutations can be realized based on RSA, but the certification proofs add an additional efficiency burden. The scheme of Neven [2008] is also based on RSA and does not need certification, but the signature size grows with the number of signers.

Use of aggregate signature schemes in blockchains

Aggregate signatures allow for a more compact representation of multiple signatures. Especially for data that is stored on many replicas, such as the blockchain itself, the overall savings in storage can be significant. One possible application is the aggregation of the signatures of all transactions within each block into a single signature.

1.3.3 MULTI-SIGNATURES

A *multi-signature scheme*, initially proposed by Itakura and Nakamura [1983], enables a group of signers to produce a joint signature on a common message. A trivial implementation of a multi-signature scheme concatenates the individual signatures of all signers; the size of such a multi-signature, however, grows linearly with the number of signers. Specialized multi-signature schemes achieve better efficiency in terms of signature size.

The security of multi-signature schemes is more subtle than that of plain digital signature schemes. For instance, malicious signers can choose their keys dependent on the public keys of other participants and make compact multi-signatures ambiguous in terms of who signed [Horster et al. 1995]. Today's security definitions for multi-signature schemes are based on the work of Micali et al. [2001]. Their scheme, however, requires an interactive setup phase during which all possible signers interact; this is clearly not appropriate for today's applications in blockchains as no further parties can join after the setup is complete.

Several multi-signature schemes allow non-interactive signing—that is, all signers produce their contributions to the signature independently and any party can combine the contributions into the final multi-signature. In these schemes, each signer produces its contribution as in a standard digital signature scheme, and there is an additional algorithm to combine all contributions into the multi-signature.

An early scheme that follows the above approach and is based on the Gap Diffie-Hellman assumption has been provided by Boldyreva et al. [2003]; the security of this scheme requires that the signers prove knowledge of their secret key by, for example, signing the certificate-signing request during key registration at a certificate authority [Ristenpart and Yilek 2007]. (This is more appropriate in the permissioned setting, where users register their keys.) A multi-signature scheme that does not require key registration can be obtained by using the aggregate signature scheme described by Bellare et al. [2007], which is based on the work of Boneh et al. [2003], as described in Section 1.3.2. Finally, Neven [2008] describes a data authentication scheme based on the RSA assumption; the gain in reduced bandwidth overhead is, however, only significant for large messages.

A recent scheme by Boneh et al. [2018] additionally supports *aggregation of public keys*: instead of requiring all signers' public keys for verifying the signature, the verifier only requires a short aggregation of those keys. This can save considerable space in the transaction data structure. The scheme of Boneh et al. [2018] is also based on BLS signatures, and can be easily described based on the scheme of Boneh et al. [2003] given in Section 1.3.2.

We again use the described bilinear pairing setting and each user uses keys $sk = x$ and $vk = g_2^x$. To user keys $vk_1, \ldots, vk_n$ for signing keys $x_1, \ldots, x_n$, we associate values $a_i \leftarrow H_1(vk_i \mid vk_1 \ldots vk_n)$, where H_1 is a hash function that maps into $\mathbb{Z}_q$, with q being the order of $\mathcal{G}_1$ and $\mathcal{G}_2$.

The keys are then aggregated as

$$avk = \prod_{i=1}^{n} vk_i^{a_i}.$$

For a given message m, each user computes their share $s_i \leftarrow H(m)^{a_i x_i}$, and the multi-signature is then computed as in Boneh et al. [2003] via $s \leftarrow \prod_{i=1}^{n} s_i$. Finally, the verification now proceeds as $e(s, g_2) = e(H(m), avk)$.

Example 1.5 (Use of multi-signature schemes in MIMO transactions)   Maxwell et al. [2019] discuss the use of multi-signatures for Bitcoin. A multi-input multi-output (MIMO) transaction in Bitcoin contains signatures (on the same data) with all keys corresponding to the inputs of the transaction. A multi-signature can therefore decrease the size of the transactions stored in the blockchain. Figure 1.4 exemplifies this.

Multi-signatures can also be used in a different way: spending of a single UTXO can be made dependent on the presence of multiple signatures relative to different verification keys. This functionality is useful if multiple persons are needed to sign off on a transaction, or if different keys are stored in different locations for security purposes. In such a setting, a multi-signature allows transactions to be more compact since the input only has to contain a single signature that is verified relative to multiple keys.

The schemes of Maxwell et al. [2019] and Boneh et al. [2018] additionally support aggregation of public keys, which informally means that several verification keys used in a multi-signature can be combined into a single, aggregated verification key. Verification is then performed with respect to this single key. This property allows more compact representation of multi-signature transactions (in particular for the output).

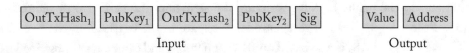

Input   Output

FIGURE 1.4: Data structure of a multi-sign UTXO transaction (simplified). The input references two outputs of previous transactions via the transactions' hashes (OutTxHash$_i$). It contains the signature public key (PubKey$_i$) and multi-signature (Sig) to allow for verification. Some schemes (e.g., Boneh et al. [2018], Maxwell et al. [2019]) additionally allow key aggregation, such that all values PubKey$_i$ are combined into a single value of the same size.

1.3.4   THRESHOLD SIGNATURES

Threshold signatures are a specific type of multi-signatures. For a set of n users and a specific threshold $t \leq n$, generation of a signature is possible if and only if t signers collaborate. While multi-signature schemes may allow different parties to generate their keys independently, threshold signature schemes require a special setup phase. Threshold cryptosystems were introduced by Desmedt and Frankel [1989]; however, their scheme required a trusted party to generate all keys and is not practical from today's perspective.

Shoup [2000] described the first practical threshold signature scheme based on RSA that has a non-interactive signing phase. The protocol requires a trusted setup phase to generate the keys; this requirement has been overcome by Damgård and Koprowski [2001]. More recently, Gennaro et al. [2008] have described a variant of the scheme that works for dynamically changing groups of users. The BLS signature scheme of Boneh et al. [2001], described in Section 1.3.1, can also be used as a threshold signature scheme, which was first observed by Boldyreva et al. [2003]. We describe the threshold version of the scheme next.

This scheme is based on a concept called secret sharing. Shamir [1979] observed that a secret number can be shared between a set of n parties such that any set of $t + 1$ parties can reconstruct the secret, but it remains hidden for any threshold of t parties, for $t < n$. The scheme uses polynomials, as a degree t polynomial is determined by any set of (at least) $t + 1$ points. To share a secret s, one chooses a random polynomial f of degree t such that $f(0) = s$. For parties $i = 1, \ldots, n$, each party i receives $f(i)$. Given those points, any set of $t + 1$ parties can compute $f(0)$ using Lagrange interpolation.

In the threshold signature scheme, the shared secret is a signing key $x \in \mathbb{Z}_q$ of a BLS signature scheme, and each user i receives a share $x_i \in \mathbb{Z}_q$. The initial distribution of keys can be achieved via the distributed key generation protocol of Gennaro et al. [2007]. To produce a contribution to a signature, a user i simply computes $s_i \leftarrow H(m)^{x_i}$, that is, a BLS signature with their key x_i. Recall that any group of $t + 1$ users could reconstruct x using Lagrange interpolation. To obtain the signature $s = H(m)^x$ from shares $s_{i_1}, \ldots, s_{i_{t+1}}$, one simply performs the same computation in the exponent of $H(m)$, using multiplication and scalar exponentiation instead of addition and scalar multiplication.

Example 1.6 (Use of threshold signature schemes in blockchains)   Gennaro et al. [2016] suggest the use of threshold DSA for generating signatures in Bitcoin transactions, as a measure against compromise of secret keys, and recent work of Lindell and Nof [2018] makes this idea more practical. A further area where threshold signatures are useful is permissioned blockchain systems, meaning systems in which all users have registered identities, and where a natural model is for a threshold of nodes to agree on a specific action.

1.3.5 FORWARD-SECURE SIGNATURES

In digital signature schemes, an attacker that obtains an honest user's signing key is able to create arbitrary messages in the name of that honest user. In the traditional, *public-key infrastructure*-based (PKI-based) Internet protocols this allows the attacker to access messages or resources on behalf of that particular user, until the user's certificate is revoked or expires. In the case of blockchain systems, particularly ones based on proof of stake, the issue is more subtle due to two main reasons: (a) the signature keys are used for two different purposes, namely spending the tokens *and* participation in consensus, and (b) the validity of the keys is not established by external means, but by the consensus itself. If an honest user's keys become known to an attacker, for instance because the user spent all tokens and stopped diligently protecting the now apparently useless signature key, the attacker can still use the key to produce messages that relate to consensus on previous blocks and maybe create a fork! This is often referred to as a *long-range attack*.

One method to prevent attacks on such "outdated" parts of the consensus is the use of forward-secure signatures, as proposed by Anderson [2002] and Bellare and Miner [1999]. In a nutshell, forward-secure signature schemes allow users to verify that a certain signature was generated at a certain point in time, so that the attacker cannot arbitrarily *backdate* the signatures. The guarantees are often achieved by *evolving* the signing key after each signature is generated and erasing the previous key. The practicality of several different schemes has been compared by Cronin et al. [2003]. A recent, efficient scheme for forward secure multi-signatures has been presented by Drijvers et al. [2020].

Use of forward-secure signature schemes in blockchains

Forward-secure signature schemes are used by proof-of-stake-based systems such as Algorand [Chen and Micali 2019] or the Ouroboros protocols (e.g., Ouroboros Praos [David et al. 2018]) that build the foundation of the Cardano blockchain. The basic idea is to have protocol participants evolve their signature keys after each use of the key during the protocol, such as during consensus. Since the "old" signature key is deleted, an attacker that obtains the participant's key after the operation will not be able to produce messages related to this or any previous consensus round.

1.4 VERIFIABLE RANDOM FUNCTIONS

A verifiable random function is a cryptographic primitive introduced by Micali et al. [1999].

Informally, a verifiable random function is a function that is instantiated by generating a pair of public and private keys. The private key is needed to evaluate the function on input of a string. The output of the evaluation of a verifiable random function on input of a string x is a pair (σ, π) with special properties. First of all, σ is indistinguishable from a random string for any probabilistic polynomial-time distinguisher that does not have access to the secret key, except that it can ask for evaluations of the function on any other string. Secondly, the string π allows an honest probabilistic polynomial-time verifier to check that σ is correct according to the public key and the input x.

The role of verifiable random functions in blockchains is mainly due to their use in consensus protocols based on honest majority of stake. Indeed, in such protocols, each player should locally check whether, according to his/her stake, (s)he is eligible or not to propose the message that can be given in output by the consensus protocol. More specifically, having in mind proof-of-stake consensus in the context of blockchains, a player evaluates a verifiable random function to check if he/she is eligible to propose a block to the peer-to-peer network. Usually eligibility is obtained by checking that σ corresponds to a very small number (i.e., all most significant bits of σ must be zeros). The proof π plays a crucial role to avoid a situation where every player could claim to be eligible by selecting a bogus σ. The unpredictability of σ is required to guarantee that players that are eligible to propose a block remain hidden until they actually announce the block. Verifiable random functions have been studied intensively, obtaining likewise efficient constructions like the one of Dodis and Yampolskiy [2005].

Use of verifiable random functions in blockchains

Examples of proof-of-stake consensus protocols that make crucial use of verifiable random functions are Algorand [Chen and Micali 2019], Ouroboros Praos [David et al. 2018], and Ouroboros Genesis [Badertscher et al. 2018].

The verifiable random function used in Ouroboros Praos makes use of two hash functions H and H', and is proven secure based on the standard computational Diffie-Hellman assumption and moreover assuming that those two functions are random oracles. The public key is a group element $v = g^x$ and x is the secret key. The evaluation of the verifiable random function on input of a message m consists of a value $y = H(m|u)$ where $u = H'(m)^x$. The proof that y is the evaluation of m consists of u and of a non-interactive zero-knowledge proof (still in the random oracle model) that the discrete logarithm of u to the base $H'(m)$ is equal to the discrete logarithm of v to the base g.

1.5   COMMITMENT SCHEMES

A commitment scheme allows a sender to encode a message satisfying the following three properties.

Hiding.   The message remains hidden to an adversarial receiver until the sender decides to reveal it.

Binding.   With overwhelming probability there is at most one message that an adversarial sender can successfully reveal.

Correctness.   Honest senders can efficiently encode and reveal a message while honest receivers can efficiently check the correctness of the revealed message.

The role of commitment schemes in blockchains is crucial when privacy is taken into account. Indeed, several potential applications of blockchains involve private data that should remain private while at the same time they must be processed in a publicly verifiable way through a blockchain. Commitment schemes can be a solution to the above tension between privacy and public verifiability since they allow an encoding of the private information to upload on a blockchain. The hiding property maintains the privacy of the information. The binding property guarantees that there is unique information to consider. Correctness allows the information to be revealed later on-chain when privacy is not an issue anymore, or off-chain when the information should still remain private to the public but can be revealed to some specific party.

Commitment schemes exist under the minimal assumption of the existence of one-way functions [Naor 1991]. More concretely, there exist efficient constructions of commitment schemes under standard number-theoretic assumptions, such as the Pedersen commitment scheme [Pedersen 1991] that is based on the discrete logarithm assumption. In this scheme there are some random elements g, h of a group $\mathcal{G}$ of sufficiently large prime order q that are selected either by the receiver or by some external trusted party. The goal is to keep secret the discrete logarithm of h to the base g. A commitment of a message m consists of selecting a random r in Z_q and computing $c = g^m \cdot h^r$. To reveal a commitment the sender can just send the pair (m, r) that can be verified by the receiver recomputing the commitment and checking equality with c. This scheme is re-randomizable (i.e., $c' = c \cdot h^s$ for a random s is again a commitment of m) and enjoys a specific homomorphic property that makes it useful in several applications: given two commitments c and c' respectively of messages m and m', then $c'' = c \cdot c'$ is a commitment of the message $m + m'$. Finally, Pedersen commitments are unconditionally hiding and thus very useful when post-quantum security is desired.

Use of commitment schemes in blockchains

The above potential benefits of commitment schemes are largely exploited by privacy-preserving blockchains like Monero [Noether and Mackenzie 2016] and Zerocash [Ben-Sasson et al. 2014a]. In particular, Zerocash guarantees the unlinkability of coin transfers by leveraging the properties of commitment schemes and zero-knowledge proofs that we will discuss in Section 1.6. More specifically, in Zerocash the serial number of a coin is committed on the blockchain and knowledge of the committed value is used by the owner of the coin to prove possession of the coin (this is one of the components that allows one to transfer a coin anonymously). Previously, Adam Back [2013] had proposed constructing "confidential transactions" using Pedersen commitments in order to add privacy to Bitcoin transactions. This was how using commitments for private transactions was first introduced.

1.6   NON-INTERACTIVE PROOFS

The classical mathematical concept of a proof consists of generating a string that can be used by any verifier to check the truthfulness of a theorem. Therefore classical proofs are non-interactive and publicly verifiable. Giving away a proof, however, can sometimes be inconvenient since it could include a secret key or some confidential information that a prover might not want to reveal.

With the purpose of allowing proof of a theorem while preserving the privacy of data required to produce a proof, Goldwasser, Micali, and Rackoff introduced the concept of interactive zero-knowledge proofs [Goldwasser et al. 1985]. In such proofs a user (the prover) manages to convince another user (the verifier) that some claim is true, without revealing any side information.

An important subtlety is the difference between proofs and arguments. In a proof system even an unbounded adversarial prover cannot convince the verifier of a false claim with non-negligible probability. In an argument system the security of the verifier is limited to polynomial-time adversarial provers. The theorem is often named the instance and the information that allows the prover to convince the verifier is often named the witness. Another subtlety is that in some applications what really matters is that the prover actually knows a witness for convincing a verifier about the truthfulness of a theorem. This motivated the notions of proofs and arguments of knowledge [Bellare and Goldreich 1992]. Notice that even though interaction allows for protecting the privacy of the prover's input, this comes at the price of losing public verifiability since only the verifier that interacts with the prover gets convinced of the veracity of the theorem.

Non-interactive zero knowledge

In many practical applications, one prover must convince many verifiers about the truthfulness of a theorem. This is a public verifiability property that has targeted the research on constructing non-interactive zero-knowledge (NIZK [Blum et al. 1988]) proofs, where once the proof is created, it can be verified by anybody. NIZK constructions require some trusted setup (e.g., the common reference string model) or heuristic security (e.g., the random oracle model). The latter model is very popular in light of the Fiat-Shamir transform [Fiat and Shamir 1986] that efficiently converts some specific (i.e., constant-round and public-coin) proofs with some weak form of zero knowledge (i.e., zero knowledge w.r.t. honest verifiers) into NIZK proofs secure even with regard to unbounded provers as long as they make only a polynomial number of queries to the random oracle.

Relying on trusted parameters seems to be in contrast with the decentralization requirements of a blockchain. The most common instantiation of trusted parameters is the so-called common reference string model, where a string with a specific distribution is somehow sampled and made available to all players. An adversarial choice of this string usually fully compromises the security of all protocols relying on it. This limitation has been recently relaxed through secure multi-party computation protocols and will be discussed later in this chapter.

SNARGs and SNARKs

A succinct non-interactive argument (SNARG) is a non-interactive argument system where the proof computed by the prover is short in length and fast to verify. The first constructions [Micali 1994, Kilian 1994] required random oracles. More recently, definitions and constructions have instead considered the existence of a common reference string available to both prover and verifier. There are several definitions in the literature aimed at formalizing what "short" and "fast" should actually mean. The most common definitions require the length of the proof and the running time of the verifier to be bounded by $p(\lambda + |x|)$, where λ is the security parameter, x is the instance, and p is a fixed polynomial (i.e., it does not depend on the specific language to which x is supposed to belong in order to represent a truthful theorem). Some more relaxed definitions allow the proof to be longer than that but sublinear in the witness size. Some more restrictive definitions require the length of the proof to depend only on the security parameter.

When a SNARG must also be an argument of knowledge, then the name SNARK is used. As proven in Gentry and Wichs [2011], SNARGs exist only under very strong ("non-falsifiable") assumptions.

Several SNARKs have also been proposed showing the zero-knowledge property (zk-SNARKs) and relying on pairing assumptions. In such recent constructions (see, e.g., Danezis et al. [2014], Gennaro et al. [2013], Groth [2016], Groth [2010], Lipmaa [2012], Lipmaa [2013]), the argument consists of a few group elements and the verifier only needs to compute a small number of exponentiations and pairings in bilinear groups.

However, efficient zk-SNARKs require a long common reference string. There exist SNARKs that are less computationally efficient but that, on the other hand, require a shorter common reference string (see, e.g., Ben-Sasson et al. [2014b] and Bitansky et al. [2013]).

STARKs

Scalable and transparent arguments of knowledge (STARKs) have been more recently introduced [Ben-Sasson et al. 2018] in order to mitigate some of the weaknesses of SNARKs. A zero-knowledge STARK does not need a common reference string and security is proven in the random oracle model, therefore following the approach of Micali [1994] and Kilian [1994] avoiding trusted parameters. Another appealing property of STARKs is that currently there is no concrete attack to STARKs even leveraging quantum computing. The price to pay for the above advantages compared to SNARKs is succinctness. Indeed, STARKs do not simply consist of a few group elements and concretely the size of a STARK easily reaches a few hundred kilobytes or even some megabytes. Currently STARKs are not yet used in blockchains but they have been proposed as possible tools in future versions of well known blockchains (e.g., Ethereum).

Use of zk-SNARKs in blockchains

SNARKs have a crucial role in some privacy-preserving cryptocurrencies [Ben-Sasson et al. 2014a]. For instance, in Zcash a zk-SNARK must be computed in order to run a privacy-preserving coin transfer.

1.7   PRIVACY-ENHANCING SIGNATURES

A *ring signature* allows a signer to sign a message such that the authenticity can be verified relative to a *ring* of users—that is, a spontaneous set of users chosen by the signer when creating the signature. The other members of the ring do not cooperate in the signing or may not even be aware of it. Ring signatures provide a certain level of anonymity since a verifier will only learn that the signer is among the users in the ring, but will not learn which user has signed. The term *ring signature* was coined by Rivest et al. [2001]. The work of Bender et al. [2009] points out that previous constructions of ring signatures assume that public keys are generated honestly; they provide a stricter security definition. They then also describe a generic construction and two more efficient constructions based on specific computational assumptions and only for rings of size 2.

Ring signatures in which the size of the signature grows with the number of ring members are relatively efficient, such as the RSA-based scheme originally proposed by Rivest et al. [2001], the scheme of Herranz and Sáez [2003] based on Schnorr signatures, or the scheme of Boneh et al. [2003] based on BLS signatures. Few schemes even achieve a signature size that is constant in the number of participants, such as that of Dodis et al. [2004] based on RSA or the recent work of Qin et al. [2018] based on the hardness of discrete logarithm problem.

Ring signatures with advanced security properties exist and are interesting for use in combination with blockchains. For instance, *threshold ring signature schemes* combine threshold signing with the anonymity guarantees of ring signatures. A first scheme has been described by Bresson et al. [2002], based on the RSA assumption and secure in the random-oracle model. Wong et al. [2003] have described a construction extending the original work of Rivest et al. [2001]. A *linkable ring signature scheme* allows determination of whether two ring signatures have been created by the same user [Liu et al. 2004, Liu and Wong 2005]; a scheme with constant-size signatures based on RSA has been proposed by Tsang and Wei [2005].

Accountable ring signatures, allowing specification of one designated opener that can revoke the anonymity, have been introduced by Xu and Yung [2004]. An implementation based on the decisional Diffie-Hellman (DDH) assumption, in which the signature size grows logarithmically with the number of ring users, has been proposed by Bootle et al. [2015]. This has been improved to constant-size signatures by Lai et al. [2016] and Kumawat and Paul [2017].

Traceable ring signatures, as introduced by Fujisaki and Suzuki [2008], allow each user to use the ring signature only once for a certain context; this can be used for preventing double spending. The original scheme is based on the DDH assumption and proven in the random-oracle model [Fujisaki and Suzuki 2008].

Use of ring signatures in blockchains

Various cryptocurrencies have used ring signatures to improve transaction privacy. The underlying idea is that a transaction is signed with a ring signature, which is signed relative to a certain number of transaction outputs, hiding which exact output was spent. Since a vanilla implementation would allow for double spending, ring signatures with additional features are used. The CryptoNote protocol [van Saberhagen 2013], for instance, is based on the traceable ring signature scheme of Fujisaki and Suzuki [2008]. The Monero cryptocurrency uses the scheme of Liu et al. [2004].

Group signatures

A *group signature*, initially proposed by Chaim and van Heist [1991], can be seen as a ring signature with a fixed set of participants: a designated *group manager* can add participants to a group or remove them, and each participant can sign a message on behalf of the group. The group manager can also revoke the anonymity of each signer. Due to the centralized role of the manager, group signatures seem less applicable in blockchain networks than their ring counterparts.

1.8   SECURE MULTI-PARTY COMPUTATION

Secure multi-party computation (MPC) [Yao 1982, Goldreich et al. 1987] is a security definition that models the security of a real-world cryptographic protocol aimed at realizing an ideal functionality. Informally, whenever players want to compute a function of their inputs, one can securely run a computation through a trusted third party (TTP). The TTP can just collect the inputs through private channels, then it could compute and send to each player the respective output. This is an ideal-world computation.

The real world replaces the TTP with a cryptographic protocol that aims at protecting the input-output privacy of players and the correctness of the computation.

MPC focuses on assessing the fact that a real-world protocol is essentially as secure as an ideal-world computation by showing that any adversarial behavior in the real world can be replicated also in the ideal world. Since the latter is secure by inspection, the real-world protocol is secure too.

Use of MPC in blockchains

The impact of MPC with blockchains is strongly related to the above issue with the (long) common reference string required by zk-SNARKs. Indeed, if the common reference string is generated adversarially, then the entire security of the zk-SNARKS could be easily violated. Therefore, one should not rely on a trusted player to generate the common reference string, since this would require contradicting the entire point of decentralizing trust through blockchains. MPC "ceremonies" have been proposed to allow many volunteers to join MPC computations in order to generate common reference strings that remain secure if at least one player correctly computes his steps in the protocol.

In 2018 a ceremony using an MPC protocol of Bowe et al. [2017] produced a partial common reference string for the pairing-based zk-SNARK scheme of Groth [2016]. The ceremony was named "Powers of Tau" and the resulting parameters can be considered trusted as long as at least one participant destroyed the randomness used during the ceremony.

BIBLIOGRAPHY

The numbers following each entry are the page numbers on which that entry is cited.

R. Anderson. Two remarks on public key cryptology. Technical Report UCAM-CL-TR-549, University of Cambridge, December 2002. 12

A. Back. Bitcoins with homomorphic value (validatable but encrypted). https://bitcointalk.org/index .php?topic=305791.0, 2013. (Accessed 20 Dec. 2019.) 14

C. Badertscher, P. Gazi, A. Kiayias, A. Russell, and V. Zikas. Ouroboros Genesis: Composable proof-of-stake blockchains with dynamic availability. In D. Lie, M. Mannan, M. Backes, and X. Wang, editors, *Proceedings of the 2018 ACM SIGSAC Conference on Computer and Communications Security, CCS 2018, Toronto, Ontario, Canada, October 15–19, 2018*, pages 913–930. ACM, 2018. 13

M. Bellare and O. Goldreich. On defining proofs of knowledge. In E. F. Brickell, editor, *Advances in Cryptology—CRYPTO '92, 12th Annual International Cryptology Conference, Santa Barbara, California, USA, August 16–20, 1992, Proceedings*, volume 740 of *Lecture Notes in Computer Science (LNCS)*, pages 390–420. Springer, 1992. 15

M. Bellare and S. K. Miner. A forward-secure digital signature scheme. In M. Wiener, editor, *Advances in Cryptology—CRYPTO*, volume 1666 of *LNCS*, pages 431–448. Springer, 1999. 12

M. Bellare, C. Namprempre, and G. Neven. Unrestricted aggregate signatures. In L. Arge, C. Cachin, T. Jurdziński, and A. Tarlecki, editors, *Automata, Languages and Programming*, volume 4596 of *LNCS*, pages 411–422. Springer, 2007. 8, 9

M. Bellare and P. Rogaway. Random oracles are practical: A paradigm for designing efficient protocols. In *Proceedings of the 1st ACM Conference on Computer and Communications Security*, pages 62–73. ACM, 1993. 3

M. Bellare and P. Rogaway. The exact security of digital signatures—how to sign with RSA and RABIN. In U. Maurer, editor, *Advances in Cryptology—EUROCRYPT*, volume 1070 of *LNCS*, pages 399–416. Springer, 1996. 6

E. Ben-Sasson, I. Bentov, Y. Horesh, and M. Riabzev. Scalable, transparent, and post-quantum secure computational integrity. *IACR Cryptology ePrint Archive*, 2018:46, 2018. 16

E. Ben-Sasson, A. Chiesa, C. Garman, M. Green, I. Miers, E. Tromer, and M. Virza. Zerocash: Decentralized anonymous payments from Bitcoin. In *IEEE Symposium on Security and Privacy, Berkeley, California, USA, May 18–21, 2014*, pages 459–474. IEEE Computer Society, 2014a. 5, 14, 16

E. Ben-Sasson, A. Chiesa, E. Tromer, and M. Virza. Scalable zero knowledge via cycles of elliptic curves. In J. A. Garay and R. Gennaro, editors, *CRYPTO (2) 2014, Santa Barbara, California, USA, August 17–21, 2014*, volume 8617 of *LNCS*, pages 276–294. Springer, 2014b. 16

A. Bender, J. Katz, and R. Morselli. Ring signatures: Stronger definitions, and constructions without random oracles. *Journal of Cryptology* 22(1):114–138, 2009. 17

D. J. Bernstein, N. Duif, T. Lange, P. Schwabe, and B.-Y. Yang. High-speed high-security signatures. In B. Preneel and T. Takagi, editors, *Cryptographic Hardware and Embedded Systems—CHES*, volume 6917 of *LNCS*, pages 124–142. Springer, 2011. 6

N. Bitansky, R. Canetti, A. Chiesa, and E. Tromer. Recursive composition and bootstrapping for SNARKs and proof-carrying data. In D. Boneh, T. Roughgarden, and J. Feigenbaum, editors, *Proceedings of the 45th Annual ACM Symposium on Theory of Computing, STOC 2013, Palo Alto, California, USA, June 1–4, 2013*, pages 241–250. ACM Press, 2013. 16

M. Blum, P. Feldman, and S. Micali. Non-interactive zero-knowledge and its applications. In *Proceedings of the 20th Annual ACM Symposium on Theory of Computing, STOC '88, New York, New York, USA, 1988*, pages 103–112. ACM, 1988. 15

A. Boldyreva. Threshold signatures, multisignatures and blind signatures based on the Gap-Diffie-Hellman-group signature scheme. In Y. G. Desmedt, editor, *Public Key Cryptography—PKC*, volume 2567 of *LNCS*, pages 31–46. Springer, 2003. 9, 11

D. Boneh, M. Drijvers, and G. Neven. Compact multi-signatures for smaller blockchains. In T. Peyrin and S. Galbraith, editors, *Advances in Cryptology—ASIACRYPT*, volume 11273 of *LNCS*, pages 435–464. Springer, 2018. 9, 10

D. Boneh and M. Franklin. Identity-based encryption from the Weil pairing. In J. Kilian, editor, *Advances in Cryptology—CRYPTO*, volume 2139 of *LNCS*, pages 213–229. Springer, 2001. 2

D. Boneh, C. Gentry, B. Lynn, and H. Shacham. Aggregate and verifiably encrypted signatures from bilinear maps. In E. Biham, editor, *Advances in Cryptology—EUROCRYPT*, volume 2656 of *LNCS*, pages 416–432. Springer, 2003. 7, 8, 9, 10, 17

D. Boneh, B. Lynn, and H. Shacham. Short signatures from the Weil pairing. In C. Boyd, editor, *Advances in Cryptology—ASIACRYPT*, volume 2248 of *LNCS*, pages 514–532. Springer, 2001. 6, 11

J. Bootle, A. Cerulli, P. Chaidos, E. Ghadafi, J. Groth, and C. Petite. Short accountable ring signatures based on DDH. In G. Pernul, P. Y. A. Ryan, and E. Weippl, editors, *Computer Security—ESORICS*, volume 9326 of *LNCS*, pages 243–365. Springer, 2015. 17

S. Bowe, A. Gabizon, and I. Miers. Scalable multi-party computation for zk-SNARK parameters in the random beacon model. *IACR Cryptology ePrint Archive* 2017:1050, 2017. 18

E. Bresson, J. Stern, and M. Szydlo. Threshold ring signatures and applications to ad-hoc groups. In M. Yung, editor, *Advances in Cryptology—CRYPTO*, volume 2442 of *LNCS*, pages 465–480. Springer, 2002. 17

R. Canetti, O. Goldreich, and S. Halevi. The random oracle methodology, revisited. *Journal of the ACM* 51(4):557–594, July 2004. 3

D. Chaim and E. van Heist. Group signatures. In D. W. Davies, editor, *Advances in Cryptology—CRYPTO*, volume 547 of *LNCS*, pages 257–265. Springer, 1991. 18

J. Chen and S. Micali. ALGORAND: A secure and efficient distributed ledger. *Theoretical Computer Science*, https://dblp.org/rec/journals/tcs/ChenM19.html?view=bibtex, 2019. 12, 13

E. Cronin, S. Jamin, T. Malkin, and P. McDaniel. On the performance, feasibility, and use of forward-secure signatures. In *Proceedings of the 10th ACM Conference on Computer and Communications Security, CCS*, pages 131–144. ACM, 2003. 12

I. Damgård and M. Koprowski. Practical threshold RSA signatures without a trusted dealer. In B. Pfitzmann, editor, *Advances in Cryptology—EUROCRYPT*, volume 2045 of *LNCS*, pages 152–165. Springer, 2001. 11

G. Danezis, C. Fournet, J. Groth, and M. Kohlweiss. Square span programs with applications to succinct NIZK arguments. In P. Sarkar and T. Iwata, editors, *Advances in Cryptology—ASIACRYPT 2014—20th International Conference on the Theory and Application of Cryptology and Information Security, Kaoshiung, Taiwan, R.O.C., December 7–11, 2014. Proceedings, Part I*, volume 8873 of *LNCS*, pages 532–550. Springer, 2014. 16

B. David, P. Gaži, A. Kiayias, and A. Russell. Ouroboros Praos: An adaptively-secure, semi-synchronous proof-of-stake blockchain. In J. B. Nielsen and V. Rijmen, editors, *Advances in Cryptology—EUROCRYPT 2018—37th Annual International Conference on the Theory and Applications of Cryptographic Techniques, Tel Aviv, Israel, April 29–May 3, 2018, Proceedings, Part II*, pages 66–98. Springer, 2018. 12, 13

Y. G. Desmedt and Y. Frankel. Threshold cryptosystems. In G. Brassard, editor, *Advances in Cryptology—CRYPTO*, volume 435 of *LNCS*, pages 307–315. Springer, 1989. 11

W. Diffie and M. Hellman. New directions in cryptography. *IEEE Transactions on Information Theory* 22(6):644–654, November 1976. 1, 5

Y. Dodis, A. Kiayias, A. Nicolosi, and V. Shoup. Anonymous identification in ad hoc groups. In C. Cachin and J. Camenisch, editors, *Advances in Cryptology—EUROCRYPT*, volume 3027 of *LNCS*, pages 609–626. Springer, 2004. 17

Y. Dodis and A. Yampolskiy. A verifiable random function with short proofs and keys. In S. Vaudenay, editor, *Public Key Cryptography—PKC 2005, 8th International Workshop on Theory and Practice in Public Key Cryptography, Les Diablerets, Switzerland, January 23–26, 2005, Proceedings*, volume 3386 of *LNCS*, pages 416–431. Springer, 2005. 13

M. Drijvers, S. Gorbunov, G. Neven, and H. Wee. Pixel: Multi-signatures for consensus. In *USENIX Security*. https://dblp.org/rec/conf/uss/Drijvers0NW20.html?view=bibtex, 2020. 12

A. Fiat and A. Shamir. How to prove yourself: Practical solutions to identification and signature problems. In A. M. Odlyzko, editor, *Advances in Cryptology—CRYPTO '86, Santa Barbara, California, USA, 1986, Proceedings*, volume 263 of *LNCS*, pages 186–194. Springer, 1986. 15

E. Fujisaki and K. Suzuki. Traceable ring signature. *IEICE Transactions*, 91-A(1):83–93, 2008. 17, 18

R. Gennaro, C. Gentry, B. Parno, and M. Raykova. Quadratic span programs and succinct NIZKs without PCPs. In T. Johansson and P. Q. Nguyen, editors, *Advances in Cryptology—EUROCRYPT 2013, 32nd Annual International Conference on the Theory and Applications of Cryptographic Techniques,*

Athens, Greece, May 26–30, 2013. Proceedings, volume 7881 of *LNCS*, pages 626–645. Springer, 2013. 16

R. Gennaro, S. Goldfeder, and A. Narayanan. Threshold-optimal DSA/ECDSA signatures and an application to Bitcoin wallet security. In M. Manulis, A.-R. Sadeghi, and S. Schneider, editors, *Applied Cryptography and Network Security—14th International Conference*, volume 9696 of *LNCS*, pages 156–174. Springer, 2016. 11

R. Gennaro, S. Halevi, H. Krawczyk, and T. Rabin. Threshold RSA for dynamic and ad-hoc groups. In N. Smart, editor, *Advances in Cryptology—EUROCRYPT*, volume 4965 of *LNCS*, pages 88–107. Springer, 2008. 11

R. Gennaro, S. Jarecki, H. Krawczyk, and T. Rabin. Secure distributed key generation for discrete-log based cryptosystems. *Journal of Cryptology* 20(1):51–83, 2007. 11

C. Gentry and D. Wichs. Separating succinct non-interactive arguments from all falsifiable assumptions. In L. Fortnow and S. P. Vadhan, editors, *Proceedings of the 43rd ACM Symposium on Theory of Computing, STOC 2011, San Jose, California, USA, June 6–8, 2011*, pages 99–108. ACM, 2011. 16

O. Goldreich, S. Micali, and A. Wigderson. How to play any mental game or A completeness theorem for protocols with honest majority. In A. V. Aho, editor, *Proceedings of the 19th Annual ACM Symposium on Theory of Computing, New York, New York, USA, 1987*, pages 218–229. ACM, 1987. 18

S. Goldwasser, S. Micali, and C. Rackoff. The knowledge complexity of interactive proof-systems (extended abstract). In R. Sedgewick, editor, *Proceedings of the 17th Annual ACM Symposium on Theory of Computing, Providence, Rhode Island, USA, May 6–8, 1985*, pages 291–304. ACM, 1985. 15

S. Goldwasser, S. Micali, and R. Rivest. A digital signature scheme secure against adaptive chosen-message attacks. *SIAM Journal of Computing* 17(2):281–308, April 1988. 5

J. Groth. Short pairing-based non-interactive zero-knowledge arguments. In M. Abe, editor, *Advances in Cryptology—ASIACRYPT 2010—16th International Conference on the Theory and Application of Cryptology and Information Security, Singapore, December 5–9, 2010. Proceedings*, volume 6477 of *LNCS*, pages 321–340. Springer, 2010. 16

J. Groth. On the size of pairing-based non-interactive arguments. In M. Fischlin and J.-S. Coron, editors, *Advances in Cryptology—EUROCRYPT 2016—35th Annual International Conference on the Theory and Applications of Cryptographic Techniques, Vienna, Austria, May 8–12, 2016, Proceedings, Part II*, volume 9666 of *LNCS*, pages 305–326. Springer, 2016. 16, 18

J. Herranz and G. Sáez. Forking lemmas for ring signature schemes. In T. Johansson and S. Maitra, editors, *Progress in Cryptology—INDOCRYPT*, volume 2904 of *LNCS*, pages 266–279. Springer, 2003. 17

P. Horster, M. Michels, and H. Petersen. Meta-multisignature schemes based on the discrete logarithm problem. In J. H. P. Eloff and S. H. von Solms, editors, *Information Security—the Next Decade*, pages 128–142. Springer, 1995. 8, 9

K. Itakura and K. Nakamura. A public-key cryptosystem suitable for digital multisignatures. Technical Report 71, NEC, 1983. 9

J. Katz and Y. Lindell. *Introduction to Modern Cryptography, Second Edition*. Chapman & Hall, 2014. 1

J. Kilian. On the complexity of bounded-interaction and noninteractive zero-knowledge proofs. In *Proceedings of the 35th Annual Symposium on Foundations of Computer Science, Santa Fe, New Mexico, USA, November 20–22, 1994*, pages 466–477. IEEE Computer Society, 1994. 15, 16

S. Kumawat and S. Paul. A new constant-size accountable ring signature scheme without random oracles. In X. Chen, D. Lin, and M. Yung, editors, *Information Security and Cryptology—13th International Conference*, volume 10726 of *LNCS*, pages 157–179. Springer, 2017. 17

W. F. R. Lai, T. Zhang, S. S. M. Chow, and D. Schröder. Efficient sanitizable signatures without random oracles. In I. Askoxylakis, S. Ioannidis, S. Katsikas, and C. Meadows, editors, *Computer Security—ESORICS*, volume 9878 of *LNCS*, pages 363–380. Springer, 2016. 17

L. Lamport. Constructing digital signatures from a one-way function. Technical Report CSL-98, SRI International, October 1979. 5

Y. Lindell and A. Nof. Fast secure multiparty ECDSA with practical distributed key generation and applications to cryptocurrency custody. In *Proceedings of the ACM Conference on Computer and Communications Security*, pages 1837–1854, 2018. 11

H. Lipmaa. Progression-free sets and sublinear pairing-based non-interactive zero-knowledge arguments. In R. Cramer, editor, *Theory of Cryptography—9th Theory of Cryptography Conference, TCC 2012, Taormina, Sicily, Italy, March 19–21, 2012. Proceedings*, volume 7194 of *LNCS*, pages 169–189. Springer, 2012. 16

H. Lipmaa. Succinct non-interactive zero knowledge arguments from span programs and linear error-correcting codes. In K. Sako and P. Sarkar, editors, *Advances in Cryptology—ASIACRYPT 2013—19th International Conference on the Theory and Application of Cryptology and Information Security, Bengaluru, India, December 1–5, 2013, Proceedings, Part I*, volume 8269 of *LNCS*, pages 41–60. Springer, 2013. 16

J. K. Liu, V. K. Wei, and D. S. Wong. Linkable spontaneous anonymous group signature for ad hoc groups. In H. Wang, J. Pieprzyk, and V. Varadharajan, editors, *Information Security and Privacy: 9th Australasian Conference*, volume 3108 of *LNCS*, pages 325–335. Springer, 2004. 17, 18

J. K. Liu and D. S. Wong. Linkable ring signatures: Security models and new schemes. In O. Gervasi, M. L. Gavrilova, V. Kumar, A. Laganà, H. P. Lee, Y. Mun, D. Taniar, and C. J. K. Tan, editors, *Computational Science and Its Applications*, volume 3481 of *LNCS*, pages 614–623. Springer, 2005. 17

A. Lysyanskaya, S. Micali, L. Reyzin, and H. Shacham. Sequential aggregate signatures from trapdoor permutations. In C. Cachin and J. Camenisch, editors, *Advances in Cryptology—EUROCRYPT*, volume 3027 of *LNCS*, pages 74–90. Springer, 2004. 8

G. Maxwell, A. Poelstra, Y. Seurin, and P. Wuille. Simple Schnorr multi-signatures with applications to Bitcoin. *Designs, Codes and Cryptography*, pages 1–26, February 2019. 10

R. C. Merkle. Secrecy, authentication, and public key systems. Technical Report 1979-1, Stanford University, June 1979. 5

R. C. Merkle. Protocols for public-key cryptosystems. In *IEEE Symposium on Security and Privacy*, pages 122–134. IEEE, 1980. 4

S. Micali. CS proofs (extended abstracts). In *Proceedings of the 35th Annual Symposium on Foundations of Computer Science, Santa Fe, New Mexico, USA, November 20–22, 1994*, pages 436–453. IEEE Computer Society, 1994. 15, 16

S. Micali, K. Ohta, and L. Reyzin. Accountable-subgroup multisignatures. In *Proceedings of the 8th ACM Conference on Computer and Communications Security*, pages 245–254. ACM, 2001. 9

S. Micali, M. O. Rabin, and S. P. Vadhan. Verifiable random functions. In *40th Annual Symposium on Foundations of Computer Science, FOCS '99, New York, NY, USA, October 17–18, 1999*, pages 120–130. IEEE Computer Society, 1999. 12

M. Naor. Bit commitment using pseudorandomness. *J. Cryptology* 4(2):151–158, 1991. 14

G. Neven. Efficient sequential aggregate signed data. In N. Smart, editor, *Advances in Cryptology—EUROCRYPT*, volume 4965 of *LNCS*, pages 52–69. Springer, 2008. 8, 9

NIST. Digital signature standard (DSS). FIPS PUB 186-4, July 2013. 6

NIST. Secure hash standard (SHS). FIPS PUB 180-4, August 2015a. 3

NIST. SHA-3 standard: Permutation-based hash and extendable-output functions. FIPS PUB 202, August 2015b. 3

S. Noether and A. Mackenzie. Ring confidential transactions. *Ledger* 1:1–18, 2016. 14

T. P. Pedersen. Non-interactive and information-theoretic secure verifiable secret sharing. In J. Feigenbaum, editor, *Advances in Cryptology—CRYPTO '91, 11th Annual International Cryptology Conference, Santa Barbara, California, USA, August 11–15, 1991, Proceedings*, volume 576 of *LNCS*, pages 129–140. Springer, 1991. 14

M.-J. Qin, Y.-L. Zhao, and Z.-J. Ma. Practical constant-size ring signature. *Journal of Computer Science and Technology* 33(3):533–541, May 2018. 17

T. Ristenpart and S. Yilek. The power of proofs-of-possession: Securing multiparty signatures against rogue-key attacks. In M. Naor, editor, *Advances in Cryptology—EUROCRYPT*, volume 4515 of *LNCS*, pages 228–245. Springer, 2007. 9

R. Rivest, A. Shamir, and L. Adleman. A method for obtaining digital signatures and public-key cryptosystems. *Journal of the ACM* 21(2):120–126, February 1978. 1, 5

R. Rivest, A. Shamir, and Y. Tauman. How to leak a secret. In C. Boyd, editor, *Advances in Cryptology—ASIACRYPT*, volume 2248 of *LNCS*, pages 552–565. Springer, 2001. 17

C. P. Schnorr. Efficient identification and signatures for smart cards. In G. Brassard, editor, *Advances in Cryptology—CRYPTO*, volume 435 of *LNCS*, pages 239–252. Springer, 1989. 6

A. Shamir. How to share a secret. *Communications of the ACM* 22(11):612–613, November 1979. 11

V. Shoup. Practical threshold signatures. In B. Preneel, editor, *Advances in Cryptology—EUROCRYPT*, volume 1807 of *LNCS*, pages 207–220. Springer, 2000. 11

P. P. Tsang and V. K. Wei. Short linkable ring signatures for e-voting, e-cash and attestation. In R. H. Deng, F. Bao, H. Pang, and J. Zhou, editors, *Information Security Practice and Experience*, volume 3439 of *LNCS*, pages 48–60. Springer, 2005. 17

N. van Saberhagen. Cryptonote v 2.0. https://cryptonote.org/whitepaper.pdf, October 2013. 18

D. S. Wong, K. Fung, J. K. Liu, and V. K. Wei. On the RS-code construction of ring signature schemes and a threshold setting of RST. In S. Qing, D. Gollmann, and J. Zhou, editors, *Information and Communications Security*, volume 2836 of *LNCS*, pages 34–46. Springer, 2003. 17

S. Xu and M. Yung. Accountable ring signatures: A smart card approach. In J.-J. Quisquater, P. Paradinas, Y. Deswarte, and A. A. E. Kalam, editors, *Smart Card Research and Advanced Applications VI*, volume 153 of *IFIPAICT*, pages 271–286. Springer, 2004. 17

A. C.-C. Yao. Protocols for secure computations (extended abstract). In *Proceedings of the 23rd Annual Symposium on Foundations of Computer Science, Chicago, Illinois, USA, 3–5 November 1982*, pages 160–164. IEEE Computer Society, 1982. 18

AUTHORS' BIOGRAPHIES

Björn Tackmann is a Senior Researcher at the DFINITY Foundation in Zürich, Switzerland. His research is on provable security of cryptographic protocols, with a focus on blockchain technology and its applications.

Ivan Visconti is a Full Professor of Computer Science in the Department of Computer Engineering (DIEM) of the University of Salerno (Italy). His research mainly focuses on the study of privacy-preserving cryptographic protocols and on securing the integrity and confidentiality of digital systems using blockchain technology.

CHAPTER 2

A Consensus Taxonomy in the Blockchain Era

Juan A. Garay, *Texas A&M University, College Station, TX, USA*
Aggelos Kiayias, *University of Edinburgh, Edinburgh, UK*

2.1   INTRODUCTION

The consensus problem—reaching agreement distributedly in the presence of faults—has been extensively studied in the literature starting with the seminal work of Shostak, Pease, and Lamport [Lamport et al. 1982, Pease et al. 1980]. The traditional setting of the problem involves parties connected by point-to-point channels, possibly using digital signatures in order to ensure the integrity of the information that is exchanged in the course of the protocol. (For a relatively recent overview of the many variants of consensus that are considered in the distributed systems literature see Cachin et al. [2011].) Tolerating "Byzantine" behavior—that is, the presence of parties that may behave arbitrarily, possibly in malicious ways—has been one of the hallmark features in the study of the problem.

Bitcoin was introduced by Nakamoto in [2008a] and [2009], with the objective of providing a payment system that is decentralized in the sense of not relying on a central authority that should be trusted for transactions to be considered as final. Expectedly, the fundamental enabling component of the Bitcoin system is a consensus mechanism that facilitates agreement on the history of transactions. Given the conflicting interests of the Bitcoin protocol participants, such a system should be resilient to Byzantine behavior, which brings us to the main contribution of Bitcoin in the context of the consensus problem; namely, a non-traditional and novel approach from the perspective of distributed computing to solve the problem in a setting that until then had not received sufficient attention.

In light of these developments, it is important to rethink the consensus problem in the blockchain era and organize the landscape that is currently being formed, acknowledging all the new directions and novel tools that have become available in the context of consensus protocol design.

One main aspect of our work is to look into the consensus problem from a modeling perspective, providing the definitions needed to understand the problem and the solutions that have been developed over the years both in the traditional and the newer blockchain settings. In the course of this, we provide a taxonomy of protocols and impossibility results that comprehensively outline what is currently known about consensus and which questions continue to remain open. Also important is to "extract" the relevant consensus question that is particular to Bitcoin, which we term "ledger consensus" (sometimes referred to as "Nakamoto consensus"), and which is an instance of the state machine replication problem that has been long-studied in distributed systems [Schneider 1990].

Consequently, in this chapter we provide precise definitions of the relevant versions of consensus that have been investigated and systematize the existing knowledge about the problem with respect to (i) the network model, (ii) trusted setup assumptions, and (iii) computational assumptions under which, and at what costs in terms of running time and communication overhead, the problem can be solved.

We emphasize that our approach is problem-centric and the results being overviewed are conceptual and fundamental in nature, with a *feasibility focus* with respect to the "resources" mentioned above. This means that in the case of classical consensus, a very active area of research in the distributed systems community, we might only mention in passing (if at all) the more recent results on practical Byzantine fault tolerance, for example. As such, our systematization complements the various other enumerative surveys of results and publications on the subject (e.g., Bano et al. [2017], Cachin et al. [2011], and Stifter et al. [2018]).

Organization of this chapter

We start in Section 2.2 by specifying a model of multi-party protocol execution and how protocols' properties will be deemed satisfied, as well as presenting the definition of (variants of) the consensus problem. We then specify the available resources and assumptions mentioned above under which the problem has been studied: network assumptions (communication primitives and synchrony) in Section 2.3, trusted setup assumptions (no setup, public-state setup, and private-state setup) in Section 2.4, and computational assumptions (one-way functions, proofs of work, random oracle, or none of these) in Section 2.5. We then overview possibility results (i.e., constructions) and impossibility results for consensus with respect to the number of parties as a function of misbehaving parties (resp., honest vs. malicious computational power), trusted setup, running time, and communication costs—both in the traditional (point-to-point communication) setting (Section 2.6) and in the Bitcoin (peer-to-peer) setting (Section 2.7).

We present ledger consensus in Section 2.8. After defining the problem, we proceed to the evaluation of existing results through a similar lens as in the case of (standard) consensus, including an adaptation to ledger consensus of the impossibility of standard consensus for dishonest majorities.

2.2   MODEL AND DEFINITIONS

2.2.1   PROTOCOL EXECUTION

In order to provide a description of protocols and their executions, it is useful to consider a formal model of computation. We choose the interactive Turing machine (ITM)-based model put forth by Canetti [2001] and Goldreich [2001]. An ITM is like a Turing machine but with the addition of an incoming and an outgoing communication tape as well as an identity tape and a "subroutine" tape. When an instance of an ITM is generated (we will henceforth call this an ITI, for interactive Turing machine *instance*), the identity tape is initialized to a specific value that remains constant throughout the instance's execution. The ITI may communicate with other ITIs by writing to its outgoing communication tape.

Let us consider a protocol Π that is modeled as an ITM. Ideally, we would like to consider the execution of this protocol in an arbitrary setting, that is, with an arbitrary set of parties and arbitrary configuration. A common way to model this in distributed cryptographic protocols is to consider that a certain program, thought of as an *adversary*, produces this configuration and therefore the properties of the protocol should hold for any possible choice of that program, potentially with some explicitly defined restrictions. The advantage of this particular modeling approach is that it obviates the need to quantify over all the details that concern the protocol (and substitutes them with a single universal quantification over all such "environments").

Suppose now that we have a protocol Π that is specified as an ITM and we would like to consider all possible executions of this protocol in the presence of an adversary $\mathcal{A}$, that is also modeled as an ITM. We capture this by specifying a pair of ITMs $(\mathcal{Z}, C)$, called the *environment* and the *control program*, respectively. The environment $\mathcal{Z}$ is given some input that may be trivial (such as a security parameter 1^κ) and is allowed to "spawn" new ITIs using the programs of Π and $\mathcal{A}$. By convention, only a single instance of $\mathcal{A}$ will be allowed. Spawning such new instances is achieved by writing a single message to its outgoing tape which is read by C. The control program is responsible for approving such spawning requests by $\mathcal{Z}$. Subsequently, all communication of the instances that are created will be routed via C, that is, C will be receiving the instances' outgoing messages and will be approving whether they can be forwarded to the receiving parties' incoming tape. Note that this may be used to simulate the existence of point-to-point channels; nevertheless, we will take a more general approach. Specifically, the control function C, will by definition only permit outgoing messages of running ITIs to be sent to the adversary $\mathcal{A}$ (with instructions for further delivery). This captures the fact that the network cannot be assumed *de facto* to be safe for the instances that are communicating during the protocol execution (see below where we explain how the adversarial influence in the network may be constrained). Beyond writing messages that are routed through $\mathcal{A}$, ITIs can also spawn additional ITIs as prescribed by the rules hardcoded in C. This enables instances of a protocol Π to invoke subroutines that can assist in its execution. These subroutines can be sub-protocols or instances of "ideal functionalities" that may be accessible by more than a single running instance.

Given those features, the above approach provides a comprehensive framework for reasoning about protocol executions. In case a polynomial-time bound is required, in the setting where a computational assumption is employed that holds only for polynomial-time bounded programs, for example—some care needs to be applied to ensure that the total execution run time of the $(\mathcal{Z}, C)$ system remains polynomial-time. This is because even if all ITIs are assumed to be polynomially bounded, the total execution run time may not be. We refer to Proposition 3 in Canetti [2001] for more details regarding enforcing an overall polynomial-time bound.

Functionalities

Next, we will need to specify the "resources" that may be available to the instances running protocol Π—for example, access to reliable point-to-point channels or a "diffuse" channel (see below). To allow for the most general way to specify such resources, we will follow the approach of describing them as "ideal functionalities" in the terminology of Canetti [2001]. In simple terms, an ideal functionality is another ITM that may interact with instances running concurrently in the protocol execution. A critical feature of ideal functionalities is that they can be spawned by ITIs running protocol Π. In such a case, the protocol Π is defined with respect to the functionality $\mathcal{F}$. The ideal functionality may interact with the adversary $\mathcal{A}$ as well as with other ITIs running the protocol. One main advantage of using the concept of an ideal functionality in our setting, is that we can capture various different communication resources that may be available to the participants running the protocol. For instance, a secure channel functionality may be spawned to transmit a message between two instances of Π that will only leak the length of the message to the adversary. As another example, a message-passing functionality may ensure that all parties are activated prior to advancing to the next communication round (see below in synchronous vs. asynchronous executions).

The ideal functionality that captures RMT (for "reliable message transmission") is presented in Figure 2.1, assuming a synchronous operation (cf. Section 2.3.2, where we discuss how the synchrony requirement can be relaxed). Given that not all parties may be required to send a message in each communication round, the functionality has to keep track of party activations and advance to the next "round" only when all parties have been given a chance to act. (Note that an activation does not necessarily imply performing any protocol tasks.)

The ideal functionality capturing the "diffuse" operation is presented in Figure 2.2, assuming again synchronous network operation (again refer to Section 2.3.2, where we discuss how the synchrony requirement can be relaxed). A salient feature of protocols running in the $\mathcal{F}_{\text{Diffuse}}$ setting is that the session id may provide just an abbreviation of the universe of parties $\mathcal{P} = \{P_1, \ldots, P_n\}$,

Functionality $\mathcal{F}_{\mathrm{RMT}}$

The functionality interacts with an adversary $\mathcal{S}$ and a set $\mathcal{P} = \{P_1, \ldots, P_n\}$ of parties. Initialize a Boolean flag $\mathsf{flag}(P_i)$ to false and a string $\mathsf{inbox}(P_i)$ to empty, for all $i = 1, \ldots, n$.

- Upon receiving (Send, sid, P_i, P_j, m) from P_i, store (Send, sid, P_i, P_j, m) and hand (Send, sid, P_i, P_j, m, mid) to $\mathcal{S}$, where mid is a unique identifier.

- Upon receiving (Activate, sid, P_i) from P_i, set $\mathsf{flag}(P_i)$ to true. If it holds that $\wedge_{i=1}^n \mathsf{flag}(P_i)$, then for all $i = 1, \ldots, n$, set $\mathsf{flag}(P_i)$ to false, and for any (Send, sid, P_i, P_j, m, mid) that is recorded as unsent, mark it as sent, and copy (Send, sid, P_i, P_j, m, mid) to $\mathsf{inbox}(P_j)$.

- Upon receiving (Deliver, sid, mid) from $\mathcal{S}$, assuming (Send, sid, P_i, P_j, m, mid) is recorded as unsent, mark it as sent, and copy (Send, sid, P_i, P_j, m) to $\mathsf{inbox}(P_j)$.

- Upon receiving (Fetch, sid, P_i) from P_i, return $\mathsf{inbox}(P_i)$ to P_i and set $\mathsf{inbox}(P_i)$ to empty.

FIGURE 2.1: The reliable message transmission ideal functionality in the synchronous setting.

Functionality $\mathcal{F}_{\mathrm{Diffuse}}$

The functionality interacts with an adversary $\mathcal{S}$ and a set $\mathcal{U}$ of parties. Initialize a subset $A \subseteq \mathcal{U}$ to $\emptyset$, a Boolean flag $\mathsf{flag}(P_i)$ to false, and a string $\mathsf{inbox}(P_i)$ to empty, for all i such that $P_i \in \mathcal{U}$.

- Upon receiving (Send, sid, P_i, m) from P_i, set $\mathsf{flag}(P_i)$ to true, store (Send, sid, P_i, m) and hand (Send, sid, P_i, m, mid) to $\mathcal{S}$, where mid is a unique identifier.

- Upon receiving (Activate, sid, P_i) from P_i, set $A = A \cup \{P_i\}$ and $\mathsf{flag}(P_i)$ to true. If it holds that $\wedge_{i \in A} \mathsf{flag}(P_i)$, then for all $i = 1, \ldots, n$, set $\mathsf{flag}(P_i)$ to false, and for any P_j, $j \in A$, and any (Send, sid, P_i, m, mid) that is recorded as unsent for P_j, mark it as sent for P_j, and copy (Send, sid, P_i, P_j, m, mid) to $\mathsf{inbox}(P_j)$.

- Upon receiving (Deliver, sid, mid, P_i', P_j) from $\mathcal{S}$ and $j \in A$, assuming (Send, sid, P_i, m, mid) is recorded as unsent for P_j, mark it as sent for P_j, and copy (Send, sid, P_i', P_j, m) to $\mathsf{inbox}(P_j)$.

- Upon receiving (Fetch, sid, P_i) from P_i, return $\mathsf{inbox}(P_i)$ to P_i and set $\mathsf{inbox}(P_i)$ to empty.

FIGURE 2.2: The peer-to-peer diffuse ideal functionality in the synchronous setting.

of which only a subset may be activated. The functionality capturing the formalization of a hash function as a random oracle is shown in Figure 2.3.

Execution of multi-party protocols

When protocol instances are spawned by $\mathcal{Z}$, they will be initialized with an identity that is available to the program's code, as well as possibly with the identities of other instances that may run concurrently (this is at the discretion of the environment program $\mathcal{Z}$). The identities themselves may be useful to the program instance, as they may be used by the instance to address them. We will use the notation $\mathsf{VIEW}_{\Pi, \mathcal{A}, \mathcal{Z}}$ to denote an *execution* of the protocol Π with an adversary $\mathcal{A}$ and an environment $\mathcal{Z}$. The

Functionality $\mathcal{F}_{RO}$

The functionality interacts with an adversary $\mathcal{S}$ and a set $\mathcal{P} = \{P_1, \ldots, P_n\}$ of parties.

- Upon receiving (Eval, sid, x) from P_i (resp. $\mathcal{S}$), return ρ to P_i (resp. $\mathcal{S}$) if $(x, \rho) \in T$. If no entry for x is in T, then choose $\rho \leftarrow \{0, 1\}^\kappa$, add (x, ρ) in T and return ρ to P_i.

FIGURE 2.3: The random oracle ideal functionality.

execution is a string that is formed by the concatenation of all messages and all ITI states at each step of the execution of the system $(\mathcal{Z}, C)$. The parties' inputs are provided by the environment $\mathcal{Z}$, which also receives the parties' outputs. Parties that receive no input from the environment remain inactive. We denote by INPUT() the input tape of each party.

We note that by adopting the computational modeling of systems of ITMs by Canetti [2001], we obviate the need to impose a strict upper bound on the number of messages that may be transmitted by the adversary in each activation. In our setting, honest parties, at the discretion of the environment, are given sufficient time to process all messages delivered by any communication functionality available to them as a resource. It follows that denial of service attacks cannot be used to the adversary's advantage in the analysis—that is, they are out of scope from our perspective of studying the consensus problem.

Properties of protocols

In our statements we will be concerned with *properties* of protocols Π. Such properties will be defined as predicates over the random variable $\text{VIEW}_{\Pi, \mathcal{A}, \mathcal{Z}}$ by quantifying over all adversaries $\mathcal{A}$ and environments $\mathcal{Z}$.

Definition 2.1   Given a predicate Q, we say that *the protocol Π satisfies property Q* provided that for all $\mathcal{A}$ and $\mathcal{Z}$, $Q(\text{VIEW}_{\Pi, \mathcal{A}, \mathcal{Z}})$ holds.

Note that in some cases, protocols may only satisfy properties with a small probability of error over all possible executions. The probability space is determined by the private coins of all participants and the functionalities they employ. In such cases, we may indicate that the protocol satisfies the property with some (small, typically negligible in a security parameter) error probability. We will only consider properties that are polynomial-time computable predicates. Our notion of execution will capture the single-session, stand-alone execution setting for protocols; hence properties will be single-session properties.

Asynchronous vs. synchronous execution

The model above is able to capture various flavors of synchrony. This is achieved by abstracting the network communication as a functionality and specifying how the adversary may interfere with message delivery. The functionality may keep track of parties' activations and, depending on the case, ensure that parties will be given a chance to act as the protocol execution advances.

Static vs. dynamic environments

In terms of protocol participants, the model we present captures both static and dynamic environments. Specifically, it is suitable for protocols that run with a fixed number of parties that should be known to all participants in advance, but it also allows protocols for which the number of participants is not known beforehand and, in fact, it may not even be known during the course of the execution. Note that in order to allow for proper ITI intercommunication, we will always assume that the *total* set of parties is known; nevertheless, only a small subset of them may be active in a particular moment during the protocol execution.

Setup assumptions

In a number of protocols, there is a need to have some pre-existing configuration (such as the knowledge of a common reference string [CRS], or a public-key infrastructure [PKI]). Such setup assumptions can also be captured as separate functionalities $\mathcal{F}$ that are available to the protocol ITIs.

Permissioned vs. permissionless networks

In the context of the consensus problem, this terminology became popular with the advent of blockchain protocols. The Bitcoin blockchain protocol is the prototypical "permissionless" protocol where read access to the ledger is unrestricted and write access (in the form of posting transactions) can be obtained by anyone that possesses BTC (which may be acquired, in principle, by anyone that is running the Bitcoin client and invests computational power in solving proofs of work). On the other hand, a permissioned protocol imposes more stringent access control on the read and write operations that are available, as well as with respect to who can participate in the protocol. Extrapolating from the terminology as applied in the ledger setting, a permissionless consensus protocol would enable any party to participate and contribute input for consideration of the other parties. With this in mind, the traditional setting of consensus is permissioned, since only specific parties are allowed to participate; on the other hand, consensus in the blockchain setting can be either permissioned or permissionless.

Cryptographic primitives

We now overview some standard cryptographic primitives, as they are employed by some of the consensus protocols. A *digital signature scheme* consists of three PPT algorithms (Gen, Sign, Verify) such that $(vk, sk) \leftarrow \mathsf{Gen}(1^\kappa)$ generates a public-key/secret-key pair; $\sigma \leftarrow \mathsf{Sign}(sk, m)$ signs a message m; and $\mathsf{Verify}(vk, m, \sigma)$ returns 1 if and only if σ is a valid signature for m given vk. A digital signature scheme is *existentially unforgeable*, if for any PPT adversary $\mathcal{A}$ that has access to a $\mathsf{Sign}(sk, \cdot)$ oracle, the event that $\mathcal{A}$ returns some (m, σ) such that $\mathsf{Verify}(vk, m, \sigma) = 1$ has measure $\mathsf{negl}(\kappa)$, where the probability is taken over the coin tosses of the algorithms, $\mathsf{negl}()$ denotes a negligible function, and κ is the security parameter. A *collision resistant (keyed) hash function family* $\{H_k\}_{k \in K}$ has the property that $H_k : \{0, 1\}^* \rightarrow \{0, 1\}^\kappa$, it is efficiently computable, and the probability to produce $x \neq y$ with $H_k(x) = H_k(y)$ given k is $\mathsf{negl}(\kappa)$. Another, less standard primitive that has been widely deployed in consensus protocol design with the advent of the Bitcoin blockchain is *proof of work* (PoW); see Section 2.5 for more information on this primitive.

2.2.2 THE CONSENSUS PROBLEM

As mentioned earlier, *consensus* (aka *Byzantine agreement*), formulated by Shostak, Pease, and Lamport [Lamport et al. 1982, Pease et al. 1980], is one of the fundamental problems in the areas of fault-tolerant distributed computing and cryptographic protocols, in particular secure multi-party computation [Ben-Or et al. 1988, Chaum et al. 1987, Goldreich et al. 1986, Yao 1982]. In the consensus problem, n parties attempt to reach agreement on a value from some fixed domain V, despite the malicious behavior of up to t of them. More specifically, every party P_i starts the consensus protocol with an initial value $v \in V$, and every run of the protocol must satisfy (except possibly for some negligible probability) the following conditions (we note that all properties below are expressible as Q predicates according to Definition 2.1).

Termination. All honest parties decide on a value.

Agreement. If two honest parties decide on v and w, respectively, then $v = w$.

Validity. If all honest parties have the same initial value v, then all honest parties decide on v.

The domain V can be arbitrary, but frequently the case $V = \{0, 1\}$ is considered, given the efficient transformation of binary agreement protocols to the multi-valued case (cf. Turpin and Coan [1984]).[1]

There exist various measures of quality of a consensus protocol: its *resiliency*, expressed as the fraction $(\frac{t}{n})$ of misbehaving parties a protocol can tolerate; its running time—worst number of rounds by which honest parties terminate; and its communication complexity—worst total number of bits/messages communicated during a protocol run.

In the consensus problem, all the parties start with an initial value. A closely related variant is the single-source version of the problem (aka the *Byzantine Generals* problem [Lamport et al.

---

[1] Refer to Section 2.6 for more efficient transformations, where in particular the longer message is only transmitted $O(n)$ times, as opposed to $O(n^2)$.

1982], or simply (reliable or secure) "broadcast"), where only a distinguished party—the *sender*—has an input. In this variant, both the Termination and Agreement conditions remain the same, and Validity becomes:

Validity. If the sender is honest and has initial value v, then all honest parties decide on v.

A stronger (albeit natural) version of the consensus problem requires that the output value be one of the honest parties' inputs, a distinction that is only important in the case of non-binary inputs. In this version, called *strong consensus* [Neiger 1994], the Validity condition becomes:

Strong Validity. If the honest parties decide on v, then v is the input of some honest party.

Note that the distinction with the standard version of the problem is only relevant in the case of non-binary inputs. Further, the resiliency bounds for this version also depend on $|V|$ (see Section 2.6).

Another way to enhance validity is to require that the output of an honest party conforms to an *external* predicate Q [Cachin et al. 2001]. In this setting, each input v is accompanied by a proof π and is supposed to satisfy $Q(v, \pi) = 1$ (for instance, π can be a digital signature on v and Q would be verifying its validity). Note that the resulting guarantee is weaker than strong validity (since it could be the case that the decision is made on an input suggested by a corrupted party), but nevertheless it can be suitable in a multi-valued setting where only externally validated inputs are admissible as outputs.

Finally, we point out that, traditionally, consensus problems have been specified as above, in a *property-based* manner. Protocols for the problem are then proven secure/correct by showing how the properties (e.g., the Agreement, Validity, and Termination conditions) are met. Nowadays, however, it is widely accepted to formulate the security of a protocol via the "trusted-party paradigm" (cf. Goldreich et al. [1986] and Goldreich [2001]), where the protocol execution is compared with an ideal process where the outputs are computed by a trusted party that sees all the inputs. A protocol is then said to securely carry out the task if running the protocol with a realistic adversary amounts to "emulating" the ideal process with the appropriate trusted party. One advantage of such a simulation-based approach is that it simultaneously captures *all* the properties that are guaranteed by the ideal world, without having to enumerate some list of desired properties. Simulation-based definitions are also useful for applying *composition theorems* (e.g., Canetti [2000], [2001]) that enable proving the security of protocols that use other protocols as subroutines, which typically would be the case for consensus and/or broadcast protocols.

The description above captures the classical definition of the consensus problem. A related and recently extensively studied version of the problem is state-machine replication or "ledger" consensus that we will treat in Section 2.8.

On the necessity of an honest majority

Regardless of the resources available to the parties in the protocol execution, an upper bound of (less than) $n/2$ can be shown for resiliency (see, for example, Fitzi [2003]). Specifically, consider a set n of parties that are equally divided with respect to their initial values between inputs 0 and 1, and an adversary that with 1/3 probability corrupts no one (case 1), with 1/3 probability corrupts the parties that have input 0 (case 2), and with 1/3 probability corrupts the parties that have input 1 (case 3). In any case, the adversarial parties follow the protocol. Observe that case 1 requires that the honest parties converge to a common output (due to Agreement), while in the other two cases the honest parties should output 0 (case 2) and 1 (case 3). However, all three cases are perfectly indistinguishable in the view of the honest parties, and as a result a logical contradiction ensues.

2.3   NETWORK ASSUMPTIONS

2.3.1   COMMUNICATION PRIMITIVES

Consensus protocols are described with respect to a network layer that enables parties to send messages to each other. An important distinction we will make is between point-to-point connectivity vs. message "diffusion" as it manifests in a peer-to-peer communication setting.

Point-to-point channels

In this setting parties are connected with pairwise reliable and authentic channels. We call that resource RMT, for *reliable message transmission* (see Figure 2.1 in Section 2.2.1). When a party sends a message, it specifies its recipient as well as the message contents, and it is guaranteed that the recipient will receive it. The recipient can identify the sender as the source of the message. In such a fixed connectivity setting, all parties are aware of the set of parties running the protocol. Full connectivity has been the standard communication setting for consensus protocols (see Lamport et al. [1982]), although sparse connectivity has also been considered (cf. Dwork et al. [1988b] and Upfal [1992]).

In terms of measuring communication costs in this model, it will be simpler for us to use the (maximum) total number of messages in a protocol run, rather than the total number of communicated bits, assuming a suitable message size. See, e.g., Fitzi [2003, chapter 3] for a detailed account of the communication complexity of consensus (and broadcast) protocols.

Peer-to-peer diffusion

This setting is motivated by peer-to-peer message transmission that happens via "gossiping"—that is, messages received by a party are passed along to the party's peers. We refer to this basic message-passing operation as "Diffuse." Message transmission is not authenticated and it does not preserve the order of messages in the views of different parties. When a message is diffused by an honest party, there is no specific recipient and it is guaranteed that all activated honest parties will receive the same message. Nonetheless, the source of the message may be "spoofed" and thus the recipient may not reliably identify the source of the message,[2] and when the sender is malicious not everyone is guaranteed to receive the same message. Contrary to the point-to-point channels setting, parties may be aware of neither the identities of the parties running the protocol nor their precise number. The ideal functionality capturing the Diffuse operation is presented in Figure 2.2.

In order to measure the total communication costs of peer-to-peer diffusion, one needs to take into account the underlying network graph. The typical deployment setting will be a sparse constant-degree graph for which it holds that the number of edges equals $O(n)$. In such a setting, each invocation of the primitive requires $O(n)$ messages to be transmitted in the network.

Relation between the communication primitives

It is easy to see that given RMT, there is a straightforward (albeit inefficient) protocol that simulates Diffuse; given a message to be diffused, the protocol using RMT will send the message to each party in the set of parties running the protocol. On the other hand, it is not hard to establish that no protocol can simulate RMT given Diffuse. The argument is as follows, and it works no matter how the protocol using Diffuse may operate. When a party A transmits a message M to party B, it is possible for the adversary in the Diffuse setting to simulate a "fake" party A that concurrently sends a message $M' \neq M$ to B. Invariably, this will result in a setting where B has to decide which is the correct message to output and will have to produce the wrong message with non-negligible probability. It follows that Diffuse is a weaker communication primitive: one would not be able to substitute Diffuse for RMT in a protocol setting.

Other models

The above models may be extended in a number of ways to capture various real world considerations in message passing. For instance, in point-to-point channels, the communication graph may change over the course of protocol execution with edges being added or removed adversarially: something that may also result in temporary network partitions. Another intermediate model between point-to-point channels and diffusion, formulated by Okun [2005a], is to have a diffusion channel with "port awareness"—that is, the setting where messages from the same source are linkable—or without port awareness, but where each party is restricted to sending one message per round (see Section 2.3.2 for the notion of round) and their total number is known. Yet another intermediate model in terms

---

[2] Note that in contrast to a sender-anonymous channel (cf. Chaum [1981]), a diffuse channel will leak the identity of the sender to the adversary.

of partial knowledge of parties and authentication has been treated, for example, in Alchieri et al. [2008], Beimel and Franklin [1999], and follow-up works.

2.3.2 SYNCHRONY

The ability of the parties to synchronize in protocol execution is an important aspect in the design of consensus protocols. Synchrony in message passing can be captured by dividing the protocol execution into rounds where parties are activated in some sequence and each one of them has the opportunity to send messages that are received by the recipients at the onset of the next round. This reflects the fact that in real world networks, messages are usually delivered in a timely fashion and thus parties can synchronize the protocol execution in discrete rounds.

A first important relaxation to the synchronous model is to allow the adversary to control the activation of parties so that it acts last in each round, having access to all messages sent by honest participants before it decides on the actions of the adversarial participants and the ordering of message delivery for the honest parties in the next round. This concept is standard in the secure multi-party computation literature [Ben-Or et al. 1988, Chaum et al. 1988, Goldreich et al. 1987] and is commonly referred to as the "rushing adversary" [Canetti 2001]. This is captured by the functionalities in Figures 2.1 and 2.2.

A second relaxation is to impose a time bound on message delivery that is not known to the protocol participants. We shall refer to this as the "partially synchronous setting" [Dwork et al. 1988a]. This setting is easy to capture by the functionalities in Figures 2.1 and 2.2 as follows: a parameter $\Delta \in \mathbb{N}$ is introduced in each functionality that determines the maximum time a message can remain "in limbo." For each message that is sent, a counter that is initially set to 0 is introduced and counts the number of rounds that have passed since its transmission (note that this concept of round is no longer a "message-passing" round). When this counter reaches Δ, the message is copied to the inbox($\cdot$) strings for the active participants.

An even weaker setting than partial synchrony is that of message transmission with eventual message delivery, where all messages between honest parties are guaranteed to be delivered but there is no specific time bound that mandates their delivery in the course of the protocol execution. This classical model in fault tolerant distributed computing is referred to as asynchronous [Fischer et al. 1985, Lynch 1996]. Again, it is easy to adapt Figures 2.1 and 2.2 to accommodate eventual delivery, following the recent formalization of this model in Coretti et al. [2016]. Note that it is proven that no deterministic consensus protocol exists in this setting [Fischer et al. 1985], and the impossibility can be overcome by randomization [Ben-Or 1983, Chor and Dwork 1989, Feldman and Micali 1997, Rabin 1983].

Finally, in the "fully asynchronous setting" (cf. Canetti [2001]), where messages may be arbitrarily delayed *or dropped*, consensus is trivially impossible.

2.4   SETUP ASSUMPTIONS

In the context of protocol design, a setup assumption refers to information that can be available to each protocol participant at the onset of the protocol. Consensus protocols are designed with respect to a number of different setup assumptions that we outline below.

2.4.1   NO SETUP

In this setting we consider protocols where parties do not utilize any setup functionality beyond the existence of the communication functionality. Note that the communication functionality may already provide some information to the participants about the environment of the protocol; nevertheless, this setting is distinguished from other more thorough setup assumptions described below. We note that in this setting it may be of interest to consider protocol executions wherein the adversary is allowed a certain amount of precomputation prior to the onset of execution that involves the honest parties.

2.4.2   PUBLIC-STATE SETUP

A public-state setup is parameterized by a probability ensemble $\mathcal{D}$. For each input size κ, the ensemble $\mathcal{D}$ specifies a probability distribution that is sampled a single time at the onset of the protocol execution to produce a string denoted by s that is of length polynomial in κ. All protocol parties, including adversarial ones, are assumed to have access to s. In this setting, the consensus protocol will be designed for a specific ensemble $\mathcal{D}$.

The concept of a public-state setup can be further relaxed in a model that has been called "sun-spots" [Canetti et al. 2007], where the ensemble is further parameterized by an index a. The definition is the same as above, but now the protocol execution will be taken for some arbitrary choice of a. Intuitively, the parameter a can be thought of as an adversarial influence in the choice of the public string s. In this setting, the consensus protocol will be designed with respect to the ensemble class $\{\mathcal{D}_a\}_a$.

2.4.3   PRIVATE-STATE SETUP

As in the public-state case, a private-state setup is parameterized by an ensemble $\mathcal{D}$. For each input size κ and number of parties n, $\mathcal{D}$ specifies a probability distribution that is sampled a single time to produce a sequence of values $(s_1, \ldots, s_n)$. The length of each value s_i is polynomial in κ. At the onset of the protocol execution, the ensemble is sampled once and each protocol participant will receive one of the values s_i following some predetermined order. The critical feature of this setting is that each party will have private access to s_i. Observe that, trivially, the setting of private-state setup subsumes the setting of public-state setup.

As in the case of a public-state setup, it is important to consider the relaxation where the ensemble $\mathcal{D}$ is parameterized by string a. As before, sampling from $\mathcal{D}_a$ will be performed from some arbitrary choice of a. It is in this sense that private-state setup has been most useful. In particular, we can use it to express the concept of a public-key infrastructure (PKI). In this setting the ensemble $\mathcal{D}$ employs a digital signature algorithm (Gen, Sign, Verify) and samples a value $(vk_i, sk_i) \leftarrow \mathsf{Gen}(1^\kappa)$ independently for each honest participant. For each participant assumed to be adversarial at the onset of the execution, its public and secret key pair is set to a predetermined value extracted from a. The private input s_i for the i-th protocol participant will be equal to $(vk_1, \ldots, vk_n, sk_i)$, thus giving access to all parties' public (verification) keys and its own private key. Other types of private setup include "correlated randomness" [Beaver 1996], where parties get correlated random strings $(r_1, r_2, \ldots, r_n)$ drawn from some predetermined distribution, which has been used to implement a random beacon [Rabin 1983].

One may consider more complicated interactive setups, such as for example the adversary choosing a somehow based on public information available about $(s_1, \ldots, s_n)$, but we will refrain from considering those here. An alternative (and subsumed by the above) formulation of a private setup includes the availability of a broadcast channel prior to the protocol execution, which enables participants to exchange shared keys [Pfitzmann and Waidner 1992].

2.5 COMPUTATIONAL ASSUMPTIONS

The assumptions used to prove the properties of consensus protocols can be divided into two broad categories. In the information-theoretic (aka "unconditional") setting, the adversary is assumed to be unbounded in terms of its computational resources. In the computational setting, on the other hand, a polynomial-time bound is assumed.

2.5.1 INFORMATION-THEORETIC SECURITY

In the information-theoretic setting, the adversarial running time is unbounded. It follows that the adversary may take arbitrary time to operate in each invocation. Note that the protocol execution may continue to proceed in synchronous rounds; nevertheless, the running time of the adversary within each round will dilate sufficiently to accommodate its complete operation. When proving the consensus properties in this setting, we can further consider two variations: perfect and statistical. When a property, Agreement for example, is perfectly satisfied, this means that in all possible executions the honest parties never disagree on their outputs. On the other hand, in the statistical variant, there will be certain executions where the honest parties are allowed to disagree. Nevertheless, these executions will have negligible density in a security parameter (in this case, n) among all executions. We observe that the statistical setting is only meaningful for a probabilistic consensus protocol, where the honest parties may be "unlucky" in their choices of coins.

2.5.2   COMPUTATIONAL SECURITY

In the computational setting, the adversarial running time (as well as the parties' running the protocol) is restricted. We distinguish the following variants.

One-way functions

A standard computational assumption is the existence of one-way functions. A one-way function is a function $f : X \to Y$ for which it holds that f is polynomial-time computable, but the probability $\mathcal{A}(1^{|x|}, f(x)) \in f^{-1}(f(x))$, for a randomly sampled x, is negligible in $|x|$ for any polynomial-time bounded program $\mathcal{A}$. One-way functions, albeit quite basic, are a powerful primitive that enables the construction of more complex cryptographic algorithms that include symmetric-key encryption, target collision-resistant hash functions, and digital signatures [Naor and Yung 1989]; the latter in particular play an important role when categorizing consensus protocols as we see below.

Proof of work

A proof of work (PoW) [Dwork and Naor 1992] is a cryptographic primitive that enables a verifier to be convinced that a certain amount of computational effort has been invested with respect to a certain context—for example, a plaintext message or a nonce that the verifier has provided. A number of properties have been identified as important for applying the primitive specifically to blockchain protocols, including amortization resistance, sampleability, fast verification, hardness against tampering and message attacks, and almost k-wise independence [Garay et al. 2017b]. Some variants of PoWs have been shown to imply one-way functions [Bitansky et al. 2016].

2.5.3   THE RANDOM ORACLE MODEL

In the previous subsections the level of security described was captured in the standard computational model where all parties are assumed to be interactive Turing machines. In many cases, including consensus protocol design, it is useful to describe properties in the random oracle model [Bellare and Rogaway 1993]. The random oracle model can be captured as an ideal functionality $\mathcal{F}_{RO}$ (see Figure 2.3 in Section 2.2.1). In a relevant adaptation of the $\mathcal{F}_{RO}$ model for the consensus setting, the access to the oracle is restricted by a quota of $q \geq 1$ queries per party per round of protocol execution [Garay et al. 2015]. This bound is also imposed on the adversary, who is assumed to control t parties. In case $t < n/2$, the execution is said to impose honest majority in terms of "computational power."

2.6   CONSENSUS IN THE POINT-TO-POINT SETTING

In the traditional network model of point-to-point reliable channels between every pair of parties, Lamport et al. [1982] first formulated the problem in the two settings described in Section 2.5: the information-theoretic setting and the computational (also called *cryptographic*, or *authenticated*) setting. As mentioned above, in the former no assumptions are made about the adversary's computational power, while the latter relies on the hardness of computational problems (such as factoring large integers or computing discrete logs), and requires a trusted setup in the form of a PKI. Depending on the setting, some of the bounds on the problems' quality measures differ. Refer to Figure 2.4 (specifically, the left subtree) as we go through the classification below.

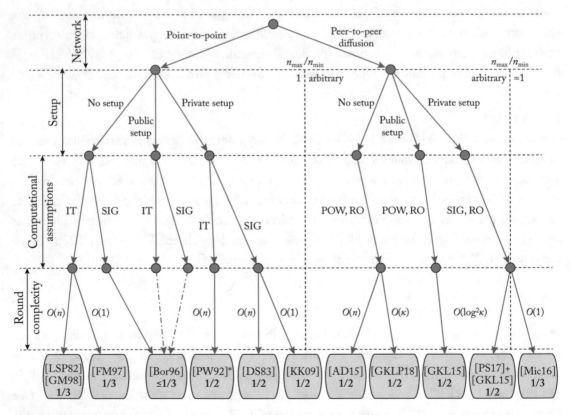

FIGURE 2.4: The taxonomy of consensus protocols and impossibility results in the synchronous setting. The dotted arrows leading to Borcherding [1996] mean that even though those cases were not explicitly considered, a similar reasoning would lead to that impossibility result. n_{max}/n_{min} refers to *participation tolerance* (cf. Section 2.7).

Number of parties

For the information-theoretic setting, $n > 3t$ is both necessary and sufficient for the problem to have a solution. The necessary condition is presented in Lamport et al. [1982] for the broadcast problem (see Fischer et al. [1986] for the consensus version of the impossibility result), as the special case of 3 parties ("generals"), having to agree on two values ('attack', 'retreat'), with one of them being dishonest. As in the information-theoretic setting (with no additional setup) the parties are not able to forward messages in an authenticated manner. It is easy to see that, as a result, an honest receiver cannot distinguish between a run where the sender is dishonest and sends conflicting messages, versus a run where a receiver is dishonest and claims to have received the opposite message, which leads to the violation of the problem's conditions (Agreement and Validity, respectively). The general case (arbitrary values of n) reduces to the 3-party case. The (broadcast) protocol presented in Lamport et al. [1982] matches this bound ($n > 3t$), and essentially consists in recursively echoing messages received in a round while excluding the messages' senders. (In the first round, only the sender sends messages.) This is done for $t + 1$ rounds, at which point the parties take the majority of the values received for that instance, returning that value as they exit that recursive step. The party's output is the value returned for the first recursive call. It was later shown that $t + 1$ rounds are optimal (see below), but the protocol requires exponential (in n) computation and communication.

Lamport et al. [1982] also formulated the problem in the computational setting, where, specifically, there is a trusted private-state setup (of a PKI), and the parties have access to a digital signature scheme. This version of the problem has been referred to as *authenticated* Byzantine agreement. In contrast to the information-theoretic setting, in the computational setting with a trusted setup the bounds for broadcast and consensus differ: $n > t$ [Lamport et al. 1982] and $n > 2t$ (e.g., Fitzi [2003]), respectively. The protocol presented in Lamport et al. [1982] runs in $t + 1$ rounds but, as in the information-theoretic setting, is also exponential-time; an efficient (polynomial-time) protocol was presented early on by Dolev and Strong [1983], which we now briefly describe. In this protocol, in the first round the sender digitally signs and sends his message to all the other parties, while in subsequent rounds parties append their signatures and forward the result. If any party ever observes valid signatures of the sender on two *different* messages, then that party forwards both signatures to all other parties and disqualifies the sender (and all parties output some default message). This simple protocol is a popular building block in the area of cryptographic protocols.

The original formulation of the problem in the computational setting assumes a PKI. In his 1996 article, Borcherding considered the situation where no PKI is available, which he refers to as "local authentication" [Borcherding 1996], meaning that no agreement on the parties' keys is provided, as each party distributes its verification key by itself. Borcherding shows that in this case, as in the information-theoretic setting above, broadcast and consensus are not possible if $n \leq 3t$, even though this setting is strictly stronger, as a dishonest party cannot forge messages sent by honest parties. The gist of the impossibility is that the adversary can always confuse honest parties about the correct protocol outcome, and digital signatures cannot help if they are not pre-associated with the parties running the protocol in advance (something only ensured given a private setup).

Regarding the "strong" version of the problem (the decision value must be one of the honest parties' input values), Fitzi and Garay [2003] showed that the problem has a solution if and only if $n > \max(3, |V|)t$ in the unconditional setting,[3] where V is the domain of input/output values, and $n > |V|t$ in the computational setting with a trusted setup, giving resiliency-optimal and polynomial-time protocols that run in $t + 1$ rounds.

Running time

Regarding the running time of consensus protocols, a lower bound of $t + 1$ rounds for deterministic protocols was established by Fischer and Lynch [1982] for the case of benign ("crash") failures, and extended by Dolev and Strong [1983] to the setting with malicious failures where messages are authenticated. As mentioned above, the original protocols by Lamport et al. already achieved this bound, but required exponential computation and communication. In contrast to the computational setting, where a polynomial-time resiliency- and round-optimal protocol was found relatively soon [Dolev and Strong 1983], in the information-theoretic setting this took quite a bit longer, and was achieved by Garay and Moses [1998]. In a nutshell, the Garay and Moses [1998] result builds on the "unraveled" version of the original protocol, presented and called *Exponential Information Gathering* by Bar-Noy et al. [1992], applying a suite of "early-stopping" (see more on this below) and fault-detection techniques to prune the tree data structure to polynomial size. Regarding strong consensus, the $t + 1$-round lower bound also applies to this version of the problem, which the protocols by Fitzi and Garay [2003] achieve (as well as being polynomial-time and resiliency-optimal).

In the $t + 1$-round lower bound for deterministic protocols, t is the maximum number of corruptions that can be tolerated in order to achieve consensus in a given model. Dolev et al. [1990] asked what the running time would be when the *actual* number of corruptions—say, f—is smaller than t, and showed a lower bound of $\min\{t + 1, f + 2\}$ for any consensus protocol, even when only crash failures occur, which is important when f is very small. They called a consensus protocol satisfying this property *early-stopping*. Faster termination, however, comes at a price of non-simultaneous termination, as they also showed that if simultaneous termination is required, then $t + 1$ rounds are necessary. (See also Dwork and Moses [1990].)

Optimal early stopping for the optimal number of parties (i.e., $n > 3t$) was achieved in the information-theoretic setting by Berman et al. [1992b]; the protocol, however, is inefficient, as it requires exponential communication and computation. Relatively recently, an efficient (polynomial-time) optimal early-stopping consensus protocol was presented by Abraham and Dolev [2015].

---

[3] The lower bound was in fact shown by Neiger [1994], who formulated this version of the problem.

The above $t+1$-round lower bound applies to deterministic protocols. A major breakthrough in fault-tolerant distributed algorithms was introduced to the field by Ben-Or [1983] and Rabin [1983], which, effectively, showed how to circumvent the above limitation by using randomization. Rabin [1983], in particular, showed that linearly resilient consensus protocols in expected *constant* rounds were possible, provided that all parties have access to a "common coin" (i.e., a common source of randomness). Essentially, the value of the coin can be adopted by the honest parties in case disagreement at any given round is detected, a process that is repeated multiple times. This line of research culminated with the work of Feldman and Micali [1997], who showed how to obtain a shared random coin with constant probability from "scratch," yielding a probabilistic consensus protocol tolerating the maximum number of misbehaving parties ($t < n/3$) that runs in an expected-constant number of rounds. Running in an expected-constant number of rounds, however, does not guarantee *simultaneous* termination [Dolev et al. 1990]. If that is required, as for example when these protocols are invoked from a higher-level protocol and all parties should move on to the next task at the same time/round (more on this below), then these common-shared-coin protocols should be run by a poly-logarithmic number of rounds.

The Feldman and Micali [1997] protocol works in the information-theoretic setting; these results were later extended to the computational setting by Katz and Koo [2009], who showed that—assuming a PKI and digital signatures—there exists an (expected-)constant-round consensus protocol tolerating $t < n/2$ corruptions. Recall that broadcast protocols in the computational setting with setup tolerate an arbitrary number (i.e., $n > t$) of dishonest parties; in contrast, the protocol in Katz and Koo [2009] assumes $n > 2t$ (as it is based on VSS—*verifiable secret sharing* [Chor et al. 1985]). Garay et al. [2007] consider the case of a dishonest majority (i.e., $n \leq 2t$), presenting an expected-constant-round protocol for $t = \frac{n}{2} + O(1)$ dishonest parties (more generally, expected $O(k^2)$ running time when $t = \frac{n}{2} + k$), and showing the impossibility of expected-constant-round broadcast protocols when $n - t = o(n)$.

As mentioned above, the speed-up on the running time of probabilistic consensus protocols comes at the cost of uncertainty, as a party that terminates can never be sure that other parties have also terminated—that is, there cannot be simultaneous termination [Dolev et al. 1990]. This is an issue when these protocols are invoked from a higher-level protocol, as a party cannot be sure how long after he receives his output from a call to such a *probabilistic termination* (PT) consensus protocol (cf. Cohen et al. [2016]) he can safely carry out the execution of the calling protocol.

The sequential composition of PT consensus protocols was addressed by Lindell et al. [2006] and the parallel composition of such protocols by Ben-Or and El-Yaniv [2003]. (The issue in the case of parallel invocations of expected-constant-round PT protocols is that the overall running time of the parallel executions is not necessarily an expected constant.) The above results on sequential and parallel composition, however, do not use simulation-based security, and it was therefore unclear how (or if) one would be able to use them to instantiate consensus (and/or broadcast) from a higher-level protocol. Such formal simulation-based (and therefore composable) definition and constructions of consensus protocols with probabilistic termination have more recently been presented in Cohen et al. [2016].

Trusted setup

We covered this aspect earlier while describing the protocols achieving the different bounds on the number of parties; here we briefly summarize it. There is no trusted setup in the unconditional setting, although in the case of randomized protocols there is the additional requirement of the point-to-point channels being private in addition to reliable, while the "authenticated" consensus protocols assume a PKI. Related to a trusted setup assumption, we remark that if a *pre-computation* phase is allowed in the information-theoretic setting where reliable broadcast is guaranteed, then as Pfitzmann and Waidner [1992] showed, broadcast and consensus are achievable with the same bounds on the number of parties as in the computational setting, using a tool known as "pseudo-signatures."

Communication cost

A lower bound of $\Omega(n^2)$ on the number of messages (in fact, $\Omega(nt)$) was shown by Dolev and Reischuk [1985] for consensus for both information-theoretic and computational security. What they showed for the latter was that the number of signatures that are required by any protocol is $\Omega(nt)$, resulting in an $\Omega(nt|\sigma|)$ bit complexity (for a constant-size domain), where $|\sigma|$ represents the maximum signature size. The first information-theoretically secure protocols to match this bound were given by Berman et al. [1992a] and independently by Coan and Welch [1989]; regarding computational security, the protocol presented by Dolev and Strong [1983] requires that many messages. By relaxing the model and allowing for a small probability of error, King and Saia [2016] presented a protocol that circumvents the impossibility result (with message complexity $\tilde{O}(n^{1.5})$).

The above bounds (except for King and Saia [2016]) reflect the fact that in typical protocols messages are communicated at least $\Omega(n^2)$ times, resulting in an overall communication complexity of at least $\Omega(\ell n^2)$ for ℓ-bit messages. Fitzi and Hirt [2006] and Hirt and Raykov [2014] showed protocols for consensus and broadcast, respectively, where the long message is communicated $O(n)$ times, which is optimal as no protocol can achieve consensus or broadcast of an ℓ-bit message with communication complexity $o(\ell n)$. See also Ganesh and Patra [2016] and Patra [2011] for further improvements.

Beyond synchrony

The case of partial synchrony, introduced in Dwork et al. [1988a], considers the existence of an unknown bound Δ that determines the maximum delay of a message that is unknown to the protocol participants.[4] As shown in Dwork et al. [1988a], the resiliency bounds presented in the point-to-point subtree of Figure 2.4 remain unaltered in the no-setup and public-setup cases, but degrade to $n/3$ in the private setup case.

In the eventual delivery setting, as mentioned above, deterministic consensus is impossible, but it is still feasible to obtain protocols with probabilistic guarantees. Furthermore, note that in this setting it is not possible to account for all of the honest parties' inputs since parties cannot afford to wait for all the parties to engage (since corrupt parties may never transmit their messages and it is impossible to set a correct time-out). In more detail, without a setup in the information-theoretic setting, it is possible to adapt the protocol in Feldman and Micali [1997] and achieve $n/4$ resilience [Feldman 1988] (see Figure 2.4). By allowing the protocol not to terminate with negligible probability, Canetti and Rabin [1993] showed how to bring the resiliency to $n/3$, which was later improved to guarantee termination with probability 1 by Abraham et al. [2008]. Efficiency improvements to the above two results (specifically, communication of the first one, and running time of the second one) were more recently presented in Bangalore et al. [2018] and Patra et al. [1980], respectively.

In the private-setup setting, assuming one-way functions, it is possible to obtain an always-terminating protocol with $n/3$ resiliency (cf. Feldman [1988]). We note that it is infeasible to go beyond $n/3$ resiliency, as shown in Ben-Or et al. [1993] and Canetti [1996], where this bound is argued for fail-stop failures, and thus the above results are optimal in this sense.

Most protocols mentioned above demonstrate the feasibility of the respective bounds. Much effort has also been dedicated to achieving practical Byzantine fault tolerance (BFT) in the eventual message delivery model. For completeness, here we mention some relevant results, with the work by Castro and Liskov [2002] as a notable instance, where they focus on a fault-tolerant replicated transactions service in the cryptographic setting with the corresponding Safety and Liveness properties (see Section 2.8), achieving $n/3$ resiliency. Cachin et al. [2005] study consensus in the same model, showing an efficient coin-tossing protocol assuming a random oracle. Other related works focusing on practical efficiency include the work by Kursawe and Shoup on "asynchronous" atomic broadcast [Kursawe and Shoup 2005] (atomicity means that broadcast executions are ordered in such a way that two broadcast requests are received in the same order by any two honest parties), following the "optimistic" approach presented in Castro and Liskov [2002] where only a "Bracha broadcast" protocol [Bracha 1984] is first attempted, reverting to the use of cryptography if things go wrong. Finally, Miller et al. [2016] improve on the communication complexity of the protocol in Cachin et al. [2001], and guarantee Liveness without any timing assumptions, which was the case in Castro and Liskov [2002].

---

[4] In Dwork et al. [1988a], partial synchrony between the clocks of the processors is also considered as a separate relaxation to the model. In the present treatment we only focus on partial synchrony with respect to message passing.

Property-based vs. simulation-based proofs

As mentioned in Section 2.2.2, consensus and broadcast protocols have been typically proven secure/correct following a *property-based* approach. It turns out, as pointed out by Hirt and Zikas [2010] (see also Garay et al. [2011]), that in the case of *adaptive* adversaries who can choose which parties to corrupt dynamically, during the course of the protocol execution (cf. Canetti et al. [1996]), most existing broadcast and consensus protocols cannot be proven secure in a simulation-based manner. The reason, at a high level, is that when the adversary (having corrupted a party) receives a message from an honest party, it can corrupt that party and make him change his message to other parties. This creates an inconsistency with the ideal process, where the party has already provided his input to the trusted party/ideal functionality that abstracts consensus. To be amenable to a simulation-based proof, instead of sending its initial message "in the clear," the sender in a broadcast protocol sends a *commitment* to the message, allowing the simulator in the ideal process to "equivocate" when the committed value becomes known in case the party has been corrupted and the initial value changed [Garay et al. 2011, Hirt and Zikas 2010].

2.7 CONSENSUS IN THE PEER-TO-PEER SETTING

Consensus in the peer-to-peer setting is the consensus problem when the available communication resource is peer-to-peer diffusion (cf. Section 2.3), a weaker communication primitive compared to point-to-point channels. (For this section, refer to the right subtree of Figure 2.4.) This setting arose with the advent of the Bitcoin blockchain protocol, and was formally studied for the first time in Garay et al. [2015]. In a nutshell, it constitutes an unauthenticated model of communication where no correlation of message sources across rounds can be established and the exact number of parties that participate may be unknown to the protocol participants. Moreover, since the adversary may inject messages into the network, an honest party cannot infer the number of participants from a message count.

We note that in a precursor model, where there is no correlation of message sources, but the point-to-point structure is still in place (albeit without authentication), Okun [2005a, 2005b] showed that deterministic consensus algorithms are impossible for even a single failure but that probabilistic consensus is still feasible by suitably adapting the protocols of Ben-Or [1983] and Feldman and Micali [1997].[5] The protocol, however, takes exponentially many rounds.

The consensus problem in the peer-to-peer setting has mostly been considered in the computational setting utilizing one-way functions and the proof-of-work (PoW) primitive (Section 2.5). The first suggestion for a solution was informally described in Aspnes et al. [2005], where it was suggested that PoWs can be used as an "identity assignment" tool, which subsequently can be used

---

[5] Hence, consensus in this setting shares a similar profile with consensus in the asynchronous network model [Fischer et al. 1985].

to bootstrap a standard consensus protocol like that in Dolev and Strong [1983]. Nevertheless, the viability of this plan was never fully analyzed until an alternative approach to the problem was informally described by Nakamoto in an email exchange [Nakamoto 2008b], where he argued that the "Byzantine Generals" problem can be solved by a blockchain/PoW approach tolerating a number of misbehaving parties strictly below $n/2$. As independently observed in Garay et al. [2014] and Miller and LaViola [2014], however, with overwhelming probability the Validity property is not satisfied by Nakamoto's informal suggestion.

The blockchain approach suggests stringing PoWs together in a hash chain and achieving agreement using a rule that favors higher concentrations of computational effort as reflected in the resulting hash chains. The inputs to the consensus problem are "entangled" within the PoWs themselves and the final output results from a processing of the hash chain. The approach was first formalized in Garay et al. [2015], where two constructions were also provided that satisfy all properties assuming a public setup.

Without access to a public setup, it is also possible to obtain a construction based on the results of Andrychowicz and Dziembowski [2015], who were the first to formalize the Aspnes et al. [2005] informal approach of using PoWs for identity assignment. Moreover, a blockchain-based approach is also possible as shown in Garay et al. [2018]. Using a private setup, it becomes feasible to use primitives such as digital signatures and verifiable random functions (by storing the public key information as part of the public part of the setup, while the secret key information is the private part of the setup) and obtain even more efficient constructions such as the consensus sub-protocol of Chen and Micali [2019].

Number of parties

One of the most important characteristics of consensus in the peer-to-peer setting is that the actual number of parties that are running the protocol is not assumed to be known in advance. Instead, the actual number of parties becomes a run-time execution parameter and the protocol is supposed to be able to tolerate a range of different possible choices for the number of parties. We capture this by posing a range of possible operational values $[n_{min}, n_{max}]$, and posit that if the actual number of parties falls within the range then the properties will be guaranteed. We call the ratio n_{max}/n_{min} for a given protocol its *participation tolerance*. We note that this notion is somewhat related to models that have been considered in fault-tolerant distributed computing and secure multi-party computation (see, e.g., Garay and Perry [1992] and Halevi et al. [2011], respectively). In such scenarios the parties are subject to two types of faults, Byzantine and benign, such as "going to sleep," but adversarially scheduled. In the latter type, the parties will cease participating in the protocol execution.

In the convention introduced in Garay et al. [2015], each party has a fixed quota of hashing queries allowed per round. As a result, the number of parties is directly proportional to the "computational power" present in the system and the total number of PoWs produced by the honest parties collectively would exceed that of the adversary, assuming honest majority with very high probability. Given this, it is tempting to imagine a direct translation of computational power to a set of identities [Aspnes et al. 2005]. The main problem is that the set of identities as perceived by the honest participants in the protocol execution might be inconsistent. This was resolved with the protocol of Andrychowicz and Dziembowski [2015], where PoWs are used to build a "graded" PKI, where keys have ranks. The graded PKI is an instance of the *graded agreement* problem [Feldman and Micali 1997], or partial consistency problem [Considine et al. 2005], where honest parties do not disagree by much, according to some metric. Subsequently, it is possible to morph this graded consistency to global consistency by running multiple instances of the protocol in Dolev and Strong [1983]. This can be used to provide a consensus protocol with resiliency $n/2$ without a trusted setup.

It is unnecessary, though, for the parties to reach consensus by establishing identities. In the first consensus protocol presented in Garay et al. [2015], the parties build a blockchain where each block contains a value that matches the input of the party that produced the block. The protocol continues for a certain number of rounds to ensure that the blockchain has grown to a certain length. In the final round, the parties remove a k-block suffix from their local blockchain, and output the majority bit from the remaining prefix. Based on the property called "common prefix" in Garay et al. [2015], it is shown that with overwhelming probability in the security parameter, the parties terminate with the same output; while, using the "chain quality" property, it is shown that if all the honest parties start with the same input, the corrupt parties cannot overturn the majority bit, which corresponds to the honest parties' input. The number of tolerated misbehaving parties in this protocol is strictly below $n/3$, a sub-optimal resiliency due to the low chain quality of the underlying blockchain protocol. The maximum resiliency that can be expected is $n/2$, something that can be shown by easily adapting the standard argument for the necessity of honest majority shown in Section 2.2.

Optimal resiliency can be reached by the second consensus protocol of Garay et al. [2015] as follows: The protocol substitutes Bitcoin transactions with a type of transactions that are *themselves* based on PoWs, and hence uses PoWs in two distinct ways: for the maintenance of the ledger and for the generation of the transactions themselves. The protocol requires special care in the way it employs PoWs since the adversary should be incapable of "shifting" work between the two PoW tasks that it faces in each round. To solve this problem, a special strategy for PoW-based protocol composition is introduced in Garay et al. [2015] called "2-for-1 PoWs." In the second solution presented in Garay et al. [2015], the number of tolerated misbehaving parties is strictly below $n/2$.

We note that all these protocols come with a hardcoded difficulty level for PoWs that is assumed to be correlated with the number of parties n. If f is the probability that at least one honest party will produce a PoW in a round of protocol execution, it holds that f approaches 0 for small n

while it approaches 1 for large n. It follows that the choice of PoW difficulty results in an operational range of values $[n_{min}, n_{max}]$ and it is possible to set the difficulty for any constant ratio n_{max}/n_{min}, so the participation tolerance of the protocol can be set to any arbitrary constant. We note that the lower bound n_{min} can be arbitrarily small as long as we are able to assume that even a single party has sufficient computational power to ensure that finding PoWs is not very rare. In case this is not true and $n < n_{min}$, the protocol cannot be guaranteed to satisfy Validity with high probability. On the other hand, if $n > n_{max}$, the protocol cannot be guaranteed to achieve agreement with high probability.

Using digital signatures and verifiable random functions (VRFs) (or just digital signatures and a hash function modeled as a random oracle), it is possible to implement the second consensus protocol in Garay et al. [2015] over an underlying blockchain protocol that uses a public-key infrastructure instead of PoWs, and allows for arbitrary participation tolerance, such as that in Pass and Shi [2017], for optimal resiliency of $n/2$. The idea is as follows: one can use VRFs for each participant to enable a random subset of elected transaction issuers in each round. The ledger will then incorporate such transactions within a window of time following the same technique and counting argument as in the second consensus protocol of Garay et al. [2015]. In Figure 2.4 this is the protocol referred to in the second leaf from the right.

Running time

In order to measure the running time that the protocols require in the peer-to-peer setting assuming PoW, one will additionally have to take into account that periods of silence—that is, rounds without any message passing—may also be required for ensuring the required properties with high probability in κ, a security parameter. In the consensus protocol derived from the protocol of Andrychowicz and Dziembowski [2015], $O(n)$ rounds are required where n is the number of parties. This can be improved to $O(\kappa)$ by, for example, using a blockchain-based approach [Garay et al. 2018]. In the public-setup setting, assuming that the number of parties fall within the operational range, the protocols of Garay et al. [2015] run in time $O(\log^2(\kappa))$.[6]

It is worth noting the contrast to the approach used in randomized solutions to the problem in the standard setting (cf. Section 2.6), where achieving consensus is reduced to (the construction of) a shared random coin, and comparable guarantees are obtained after a poly-logarithmic number of rounds in the number of parties. The probabilistic aspect in the blockchain setting stems from the parties' likelihood of being able to provide proofs of work.

---

[6] In the original formulation, the protocols in Garay et al. [2015] take $O(\kappa)$ rounds; it turns out that $O(\log^2(\kappa))$ is sufficient for the properties to be satisfied except with negligible probability.

In the private-setup setting it is possible to improve the running time to an expected constant—for example, by deploying the consensus sub-protocol of Algorand [Chen and Micali 2019] for 1/3 resiliency.

Trusted setup

The relevant trusted setup assumption in the above protocols includes a fresh random string that can be incorporated as part of a "genesis block" in the blockchain protocol setting, or in general as part of the PoWs.[7] The objective of this public setup is to prevent a pre-computation attack by the adversary that will violate the relative superiority of honest parties which would be derived by the honest majority assumption. Note that protocols that require no trusted setup such as Andrychowicz and Dziembowski [2015] and Garay et al. [2018] take advantage of a special randomness exchange phase prior to PoW calculation that guarantees freshness without the need of a common random string.

It is worth emphasizing the fundamental advantage of the PoW setting compared to other computational assumptions that have been used for consensus. Specifically, it is known that without a private setup, consensus is not possible with more than $n/3$ corruptions [Borcherding 1996], even assuming digital signatures. The $n/3$ impossibility result does not apply here since, essentially, proofs of work can make it infeasible for the adversary to present diverging protocol transcripts without investing effort for distinct PoW calculations.

Another observation is that, assuming a private setup in the peer-to-peer setting, one can simulate point-to-point connectivity and thus run any consensus protocol from the previous section. Nevertheless, this reduction is not efficient and in the peer-to-peer setting with private setup one can still obtain protocols that are more efficient (e.g., with subquadratic communication complexity).

Communication cost

The total number of transmitted messages in the consensus protocols described above is, in expectation, $O(n^2\kappa)$ for the case of Andrychowicz and Dziembowski [2015] and Garay et al. [2018], counting each invocation of the diffuse channel as costing $O(n)$ messages. For the two protocols of Garay et al. [2015], the number of messages is $O(n\kappa)$ in the public-setup setting. In the private-setup setting it can be possible to reduce this further, using techniques from Chen and Micali [2019].

We recall that an important difference with randomized consensus protocols in the standard setting is that parties send messages in every round, while in the PoW setting (honest) parties only communicate whenever they are able to produce a proof of work; otherwise, they remain silent. This

---

[7] Alternatively, the protocols would consider as valid any chain that extends the empty chain, and where the adversary is not allowed any pre-computation.

also suggests that there may be honest parties that never diffuse a message[8] and thus it is feasible to drop communication costs to below n^2 (with a probabilistic guarantee; cf. Section 2.6).

Beyond synchrony

The consensus protocols of Garay et al. [2015] in Figure 2.4 can be analyzed in the partially synchronous setting as well (refer to the full version of Garay et al. [2014] as a starting point). Recall that the way the protocols operate in this setting is that a hardcoded parameterization of difficulty provides a reasonable PoW production rate over message-passing time. The security of the protocols will then be at the theoretical maximum in terms of resiliency as long as the original estimate is close to being safe (network delay is low) but will degrade if the estimate is worse, dissipating entirely when the delay gets larger. (For the full argument, see Pass et al. [2017], which shows how the blockchain protocol's consistency collapses when delay is arbitrarily large.)

Property-based vs. simulation-based proofs

To our knowledge, there is no simulation-based treatment of consensus in the peer-to-peer setting. However, it is easy to infer a functionality abstracting the problem. The only essential difference is that the actual number of parties involved in the execution are to be decided on the fly and will be unknown to the protocol participants.

2.8   LEDGER CONSENSUS

Ledger consensus (aka "Nakamoto consensus") is the problem where a set of servers (or nodes) operate continuously accepting inputs ("transactions") that belong to a set $\mathbb{T}$ and incorporating them into a public data structure called the *ledger*. We assume that the language of all valid ledgers $\mathbb{L}$ has an efficient membership test, and moreover for all tx there is an $\mathcal{L} \in \mathbb{L}$ such that tx $\in \mathcal{L}$. We call a language $\mathbb{L}$ *trivial* if it holds that for all $tx_1, tx_2 \in \mathbb{T}$ there exists $\mathcal{L} \in \mathbb{L}$ that contains both tx_1 and tx_2. The purpose of ledger consensus is to provide a unique view of the ledger to anyone asking to see it. The ledger view of a party P is denoted by $\widetilde{\mathcal{L}}_P$ while the "settled" portion of the ledger in the view of P is denoted by $\mathcal{L}_P$. Note that it always holds $\mathcal{L}_P \preceq \widetilde{\mathcal{L}}_P$, where $\preceq$ denotes the standard prefix operation. The properties that a ledger consensus protocol must satisfy are as follows:

Consistency (or Persistence).   This property mandates that if a client queries an honest node's ledger at round r_1 and receives the response $\widetilde{\mathcal{L}}_1$, then a client querying an honest node's ledger at round $r_2 \geq r_1$ will receive a response $\widetilde{\mathcal{L}}_2$ that satisfies $\mathcal{L}_1 \preceq \widetilde{\mathcal{L}}_2$.

---

[8] Note the similarity with standard consensus in the eventual-delivery setting (Section 2.6), where not all honest parties' inputs may be accounted for.

Liveness. If a transaction tx is given as input to all honest nodes at a round r and it holds that tx is valid w.r.t. $\widetilde{\mathcal{L}}_P$ for every honest party P, then at round $r + u$ it holds that $\mathcal{L}_P$ includes tx for any honest party P.

In classical distributed systems literature, this problem is often referred to as *state machine replication* [Schneider 1990]. Consistency ensures that parties have the same view of the log of transactions, while Liveness ensures the quick incorporation of transactions. Furthermore, a third property, called "order" in Schneider [1990], is introduced which, in our notation, can be expressed as follows:

Serializability. For transactions tx, tx′, if tx is given as input to all honest nodes at a round r and it holds that tx is valid w.r.t. $\widetilde{\mathcal{L}}_P$ and tx′ $\notin \widetilde{\mathcal{L}}_P$ for every honest party P, then it holds that for any $r' > r$, the ledger $\mathcal{L}_P$ of any honest party cannot include tx′, tx in this order.

Given a consensus protocol, it is tempting to apply it in sequential composition in order to solve ledger consensus. The reduction indeed holds, but some special care is needed. First, let us consider the case where no setup is available. The construction in the synchronous network model is as follows. First, suppose that we have at our disposal a consensus protocol that satisfies Agreement, (Strong) Validity, and Termination after u rounds. The protocol has all nodes collect transactions and then run the consensus protocol with the set of transactions as their input. When the protocol terminates after u rounds, the nodes assign an index to the output (call it the i-th entry to the ledger) and move on to the next consensus instance. It is easy to see that Consistency is satisfied because of Agreement, while Liveness is satisfied with parameter u because of Strong Validity and Termination. It is worth noting that "plain" Validity by itself is not enough, since a ledger protocol is supposed to run for any given set of transactions, and as a result it is possible that no two honest nodes would ever agree on a set of inputs. In this case, Validity might just provide that honest parties agree on an adversarial value, which might be the empty string. As a result the ledger would be empty and Liveness would be violated. However, it is possible to deal with this problem without resorting to the full power of Strong Validity. For instance, it is sufficient to consider a variant of consensus where each party has an input set X_i and the joint output set S satisfies that $X_i \subseteq S$. We note that such a "union" consensus protocol can be implied by Interactive Consistency, as defined in Pease et al. [1980], and it has also more recently been considered explicitly as a consensus variant [Dold and Grothoff 2017]. Other intermediate notions of Validity such as a *predicate-based* notion [Cachin et al. 2001] can be useful here as well.

Let us now comment on how the reduction can be performed under different setup and network assumptions. First, if a setup assumption is used, observe that the above reduction will require the availability of the setup every u rounds. Given this might be impractical, one may consider how to emulate the sequence of setups using a single initial setup. This approach is non-black-box on the underlying protocol and may not be straightforward. For instance, when sequentially composing a PoW-based consensus protocol that relies on a public setup, the security of the protocol may non-trivially rely on the unpredictability of the i-th setup. Techniques related to sequential composition of a basic building block protocol have appeared in a number of ledger protocols, including Bentov et al. [2016], Chen and Micali [2019], and Kiayias et al. [2017]. Regarding network aspects, we observe that the reduction can proceed in essentially the same way in the peer-to-peer setting as in the point-to-point setting. Finally, note that when simultaneous termination is not available in the underlying consensus protocol, special care is needed in applying composition (cf. Cohen et al. [2016]).

Ledger consensus was brought forth as an objective of the Bitcoin blockchain protocol. For this reason, in the remainder of the chapter, we only consider the problem in the peer-to-peer setting, although we note that in the point-to-point setting it is possible to adapt standard BFT methods to solve the problem (see Golan-Gueta et al. [2018] and Miller et al. [2016] for some examples). We remark also that combining private setup and the peer-to-peer setting, it is straightforward to simulate the point-to-point setting by relying on the authentication information that can be made available by the setup. A pictorial overview of our protocol classification is presented in Figure 2.5.

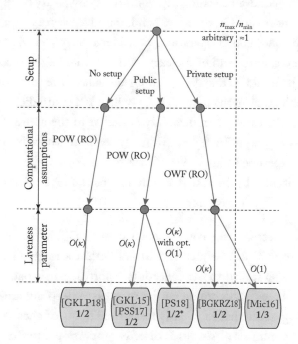

FIGURE 2.5: The taxonomy of ledger consensus protocols (peer-to-peer setting).

Number of parties

We start with an adaptation of the impossibility result for dishonest majority as shown in Fitzi [2003]. The result shows that in all the relevant cases for practice—specifically, ledger consensus with non-trivial ledgers, or providing serializability as defined above—honest majority is a necessary requirement.

Theorem 2.2 *Suppose that the transaction set $\mathbb{T}$ satisfies $|\mathbb{T}| \geq 2$. Ledger consensus is impossible in case the adversary controls $n/2$ nodes, assuming either (i) the language $\mathbb{L}$ is non-trivial or (ii) Serializability holds.*

Proof. For simplicity we describe the impossibility result in a setting where the properties are perfectly satisfied. The same argument can be easily extended to the setting where the properties are satisfied with overwhelming probability. Suppose all parties are split into two sets A_1, A_2 of size exactly $n/2$. We describe an environment and an adversary. The environment prepares two transactions $tx_1, tx_2 \in \mathbb{T}$ that are in conflict—that is, it holds that no valid $\mathcal{L}$ exists for which it holds that $tx, tx' \in \mathcal{L}$ but they can both be validly added to some ledger since they are members of $\mathbb{T}$. The environment provides at round 1 the appropriate sequence of transactions so that parties in A_b receive transaction tx_b, respectively, and advances the execution for at least u rounds, the Liveness parameter. We consider three adversaries $\mathcal{A}_0, \mathcal{A}_1, \mathcal{A}_2$. The $\mathcal{A}_0$ adversary corrupts no party and allows the execution to advance normally. On the other hand, the adversary $\mathcal{A}_b$ with $b \in \{1, 2\}$ corrupts the set of parties A_b and simulates honest operation. Consider a party $P_1 \in A_1$ and a party $P_2 \in A_2$. In case $b \in \{1, 2\}$, by Liveness, at the end of the execution it should be the case that $tx_b \in \mathcal{L}_b$. In case $b = 0$, by Consistency, it should be that $\mathcal{L}_1 \preceq \widetilde{\mathcal{L}_2}$. Given that in the three cases the executions are perfectly indistinguishable, this means that $tx_1 \in \widetilde{\mathcal{L}_2}$, which is a contradiction since $tx_2 \in \mathcal{L}_2 \preceq \widetilde{\mathcal{L}_2}$.

The argument for the case of Serializability is similar to the above. In this case, we simply assume that transactions $tx_1, tx_2 \in \mathbb{T}$ are just distinct (they do not have to be in conflict). Observe that by Liveness in the experiments above we will have that $tx_b \in \mathcal{L}_b$ for party P_b. Moreover, due to Serializability, for P_b it must be the case that transactions cannot be in the order tx_{3-b}, tx_b. This leads to a contradiction due to Consistency. □

As in the case of peer-to-peer consensus (Section 2.7), the actual number of parties n is not known in advance and may be assumed to fall within a range of operational parameters $n \in [n_{min}, n_{max}]$. This is also related to the concept of "sporadic participation" that was considered in Pass and Shi [2017], where certain honest parties may "go to sleep" for arbitrary amounts of time.

In the PoW setting, recall that each party has a fixed quota of queries to a hash function per unit of time and thus the number of parties is directly proportional to the total computational or hashing power that is available. In this setting, first Garay et al. [2015] showed that ledger consensus can be achieved when the number of corrupted parties is strictly below $n/2$. This bound was also preserved in the partially synchronous setting, as shown by Pass et al. [2017].

These results refer to a static setting where there are no large deviations in the number of parties throughout the execution. The setting where the population of parties running the protocol can dynamically (and quite drastically) change over time with the environment introducing new parties and deactivating parties that have participated was considered for ledger consensus for the first time in Garay et al. [2017a]. Their main result is that ledger consensus can be achieved in the PoW setting, assuming an honest majority appropriately restated by considering the number of parties as they change over time: Assuming n_i are the active parties at time unit i, it holds that the number of adversarial parties is bounded away from $n_i/2$.

Assuming a private setup and a setting where the adversary gets t Byzantine corruptions and s asleep parties, in Pass and Shi [2017] it is shown that ledger consensus can be achieved as long as t is strictly bounded by $a/2$ where $a = n - s$ is the number of "alert" parties; that is, the number of asleep parties may be larger than $n/2$ and hence an arbitrary participation ratio can also be achieved in this setting without resorting to PoWs. With respect to lower bounds, in the case of sleep corruptions the bound can be generalized to $a/2$; see Pass and Shi [2017]. A dynamic setting of parties was also considered in Bentov et al. [2016], David et al. [2018], and Kiayias et al. [2017] providing a similar type of results assuming a PKI with honest "stake" majority. An important deficiency shared by these works is that new parties have to be chaperoned into the system by receiving advice consistent with the views of the honest parties. This was highlighted as the "bootstrapping from genesis" problem in Badertscher et al. [2018], which resolved it via a suitable chain selection rule; in the same work, a more refined model of dynamic participation was put forth, called *dynamic availability*. This model allows finer control from the environment's perspective in terms of disconnecting parties, or having parties lose access to resources such as the clock or the hash function.

Finally, in terms of participation tolerance, we observe that an arbitrary $n_{\max}/n_{\min}$ can be achieved by protocols such as Badertscher et al. [2018] and Pass and Shi [2017], while Algorand [Chen and Micali 2019] requires $n_{\max}/n_{\min}$ to be (approximately) 1 since the expected participation is a hardcoded value in the protocol. (It is worth noting that despite this limitation, Algorand still qualifies as a peer-to-peer protocol, since the identities of the parties engaging in the protocol need not be known in advance.)

Transaction processing time

Contrary to a consensus protocol, a ledger consensus protocol is a protocol that is supposed to be running over an arbitrary, potentially long period of time. Thus, the relevant measure in this context is the amount of time that it takes for the system to insert a transaction in the log that is maintained by the participants. This relates to the parameter u introduced as part of the Liveness property, which determines the number of rounds needed in the execution model for a transaction to be included in the log. Observe that Liveness is only provided for transactions that are produced by honest participants or are otherwise unambiguously provided to the honest parties running the protocol.

In this setting we observe that Garay et al. [2015] achieves ledger consensus with processing time $O(\kappa)$ rounds of interaction, where κ is the security parameter. This result is replicated in the partially synchronous setting, where processing time takes $O(\kappa\Delta)$ rounds, and where Δ is the maximum delay imposed on message transmission. The above results assume the adversarial bounds consistent with honest majority, which are tight (cf. Pass and Shi [2018]). Considering a weaker adversarial setting, it is possible to improve Liveness; for instance, Algorand [Chen and Micali 2019] achieves an expected-constant number of rounds while Thunderella [Pass and Shi 2018] shows that the processing time can be dropped to $O(1)$ rounds worst-case, assuming an honest super-majority (i.e., adversarially controlled number of parties strictly below $n/4$) and the existence of a specific honest party called the *accelerator*.

Trusted setup

Ledger consensus can be achieved in the public- or private-state setup setting. Protocols falling into the former category are Garay et al. [2015], Garay et al. [2017a], and Pass et al. [2017], whereas protocols consistent with the latter are Bentov et al. [2016], Chen and Micali [2019], Gilad et al. [2017], Kiayias et al. [2017], and Pass and Shi [2017]. In the absence of a trusted setup, it has been shown that it is possible to "bootstrap" a ledger consensus protocol from "scratch," either directly [Garay et al. 2018] or via setting up a public-key directory using proofs of work [Andrychowicz and Dziembowski 2015]. An important further consideration between public and private setup is that in the peer-to-peer setting, the former represents what typically is consistent with the so-called permissionless setting, while the latter is consistent with the permissioned setting. This follows from the fact that anyone that has access to the peer-to-peer channel is free to participate in the protocol, if no setup or a public setup is assumed. On the other hand, in the private-setup setting, a higher level of permissioning is implied: The parties that are eligible to run the protocol need to get authorized either by the setup functionality so that they receive the private information that is related to the protocol execution, or, alternatively, interact with the parties that are already part of the protocol execution so they can be inducted. Note that the point-to-point setting is—by definition—permissioned via access to the RMT functionality.

Communication cost

Given that ledger consensus is an ongoing protocol that processes incoming transactions, defining communication costs requires some care. To our knowledge, no formal definitions of communication costs for ledger consensus have been proposed. A first approach to the problem is to consider a type of "communication overhead" on top of the transactions that are transmitted in the system. It follows that the minimum communication necessary for each bit of transaction transmitted is the diffusion of this bit. Given the above, the communication costs of ledger consensus protocols based on blockchain protocols can be seen to be constant in the sense that parties transmit, up to constant factors, more data.

Beyond synchrony
Initial work in ledger consensus protocols in the public-setup [Garay et al. 2015, Garay et al. 2017a] and the no-setup setting [Garay et al. 2016, Garay et al. 2018] assumed a rushing adversary and synchronous operation. This can be extended to the partial synchrony setting as shown in Pass et al. [2017] as well as in the full version of Garay et al. [2014] with the same limitations explained in Section 2.7.

Property-based vs. simulation-based proofs
The first simulation-based definition of ledger consensus was presented by Badertscher et al. [2017]. A refinement of this definition was presented in Badertscher et al. [2018], where it was also shown how to adapt it in a setting where a private setup is available. In terms of composability, an (expected) disadvantage for PoW-based protocols highlighted in the work of Badertscher et al. [2017] is that access to the random oracle should be specific to the current ledger protocol session.

ACKNOWLEDGMENTS

The second author was supported by H2020 Project PRIViLEDGE #780477. The authors are grateful to Christian Cachin, Arpita Patra, Björn Tackmann, and Ivan Visconti for helpful comments and suggestions.

BIBLIOGRAPHY

I. Abraham and D. Dolev. Byzantine agreement with optimal early stopping, optimal resilience and polynomial complexity. In R. A. Servedio and R. Rubinfeld, editors, *Proceedings of the 47th Annual ACM Symposium on Theory of Computing, STOC 2015, Portland, OR, USA, June 14–17, 2015*, pages 605–614. ACM, 2015. 44

I. Abraham, D. Dolev, and J. Y. Halpern. An almost-surely terminating polynomial protocol for asynchronous Byzantine agreement with optimal resilience. In R. A. Bazzi and B. Patt-Shamir, editors, *Proceedings of the 27th Annual ACM Symposium on Principles of Distributed Computing, PODC 2008, Toronto, Canada, August 18–21, 2008*, pages 405–414. ACM, 2008. 47

E. A. P. Alchieri, A. N. Bessani, J. da Silva Fraga, and F. Greve. Byzantine consensus with unknown participants. In T. P. Baker, A. Bui, and S. Tixeuil, editors, *Principles of Distributed Systems, 12th*

International Conference, OPODIS 2008, Luxor, Egypt, December 15–18, 2008. Proceedings, volume 5401 of *Lecture Notes in Computer Science (LNCS)*, pages 22–40. Springer, 2008. 38

M. Andrychowicz and S. Dziembowski. PoW-based distributed cryptography with no trusted setup. In *Advances in Cryptology—CRYPTO 2015—35th Annual Cryptology Conference, Santa Barbara, CA, USA, August 16–20, 2015, Proceedings, Part II*, pages 379–399, 2015. 49, 50, 51, 52, 58

J. Aspnes, C. Jackson, and A. Krishnamurthy. Exposing computationally-challenged Byzantine impostors. Technical Report YALEU/DCS/TR-1332, Yale University Department of Computer Science, July 2005. 48, 49, 50

C. Badertscher, P. Gazi, A. Kiayias, A. Russell, and V. Zikas. Ouroboros Genesis: Composable proof-of-stake blockchains with dynamic availability. In D. Lie, M. Mannan, M. Backes, and X. Wang, editors, *Proceedings of the 2018 ACM SIGSAC Conference on Computer and Communications Security, CCS 2018, Toronto, ON, Canada, October 15–19, 2018*, pages 913–930. ACM, 2018. 57, 59

C. Badertscher, U. Maurer, D. Tschudi, and V. Zikas. Bitcoin as a transaction ledger: A composable treatment. In Katz and Shacham [2017], pages 324–356. 59

L. Bangalore, A. Choudhury, and A. Patra. Almost-surely terminating asynchronous Byzantine agreement revisited. In C. Newport and I. Keidar, editors, *Proceedings of the 2018 ACM Symposium on Principles of Distributed Computing, PODC 2018, Egham, United Kingdom, July 23–27, 2018*, pages 295–304. ACM, 2018. 47

S. Bano, A. Sonnino, M. Al-Bassam, S. Azouvi, P. McCorry, S. Meiklejohn, and G. Danezis. Consensus in the age of blockchains. *CoRR*, abs/1711.03936, 2017. 28

A. Bar-Noy, D. Dolev, C. Dwork, and H. R. Strong. Shifting gears: Changing algorithms on the fly to expedite Byzantine agreement. *Inf. Comput.* 97(2):205–233, 1992. 44

D. Beaver. Correlated pseudorandomness and the complexity of private computations. In Miller [1996], pages 479–488. 40

A. Beimel and M. K. Franklin. Reliable communication over partially authenticated networks. *Theor. Comput. Sci.* 220(1):185–210, 1999. 38

M. Bellare and P. Rogaway. Random oracles are practical: A paradigm for designing efficient protocols. In *CCS '93, Proceedings of the 1st ACM Conference on Computer and Communications Security, Fairfax, Virginia, USA, November 3–5, 1993*, pages 62–73, 1993. 41

M. Ben-Or. Another advantage of free choice: Completely asynchronous agreement protocols (extended abstract). In R. L. Probert, N. A. Lynch, and N. Santoro, editors, *Proceedings of the 1983 ACM Symposium on Principles of Distributed Computing, PODC 1983*, pages 27–30. ACM, 1983. 38, 45, 48

M. Ben-Or, R. Canetti, and O. Goldreich. Asynchronous secure computation. In Kosaraju et al. [1993], pages 52–61. 47

M. Ben-Or and R. El-Yaniv. Resilient-optimal interactive consistency in constant time. *Distributed Computing* 16(4):249–262, 2003. 45

M. Ben-Or, S. Goldwasser, and A. Wigderson. Completeness theorems for non-cryptographic fault-tolerant distributed computation (extended abstract). In *STOC '88, Proceedings of the 28th Annual ACM Symposium on Theory of Computing*, pages 1–10, 1988. 34, 38

I. Bentov, R. Pass, and E. Shi. Snow White: Provably secure proofs of stake. *IACR Cryptology ePrint Archive*, 2016:919, 2016. 55, 57, 58

P. Berman, J. A. Garay, and K. J. Perry. Bit optimal distributed consensus. In *Computer Science Research*, pages 313–321. Springer US, Boston, MA, 1992a. 46

P. Berman, J. A. Garay, and K. J. Perry. Optimal early stopping in distributed consensus (extended abstract). In *Distributed Algorithms, 6th International Workshop, WDAG '92, Haifa, Israel, November 2–4, 1992, Proceedings*, pages 221–237. Springer Verlag, 1992b. 44

N. Bitansky, S. Goldwasser, A. Jain, O. Paneth, V. Vaikuntanathan, and B. Waters. Time-lock puzzles from randomized encodings. In M. Sudan, editor, *Proceedings of the 2016 ACM Conference on Innovations in Theoretical Computer Science, Cambridge, MA, USA, January 14–16, 2016*, pages 345–356. ACM, 2016. 41

M. Borcherding. Levels of authentication in distributed agreement. In *Distributed Algorithms, 10th International Workshop, WDAG '96, Bologna, Italy, October 9–11, 1996, Proceedings*, pages 40–55. Springer Verlag, 1996. 42, 43, 52

G. Bracha. An asynchronous [(n − 1)/3]-resilient consensus protocol. In T. Kameda, J. Misra, J. G. Peters, and N. Santoro, editors, *Proceedings of the Third Annual ACM Symposium on Principles of Distributed Computing, Vancouver, B. C., Canada, August 27–29, 1984*, pages 154–162. ACM, 1984. 47

C. Cachin, R. Guerraoui, and L. Rodrigues. *Introduction to Reliable and Secure Distributed Programming*, 2nd edition. Springer Publishing Company, Incorporated, 2011. 27, 28

C. Cachin, K. Kursawe, F. Petzold, and V. Shoup. Secure and efficient asynchronous broadcast protocols. In *Advances in Cryptology—CRYPTO 2001, 21st Annual International Cryptology Conference, Santa Barbara, California, USA, August 19–23, 2001, Proceedings*, pages 524–541. Springer, 2001. 35, 47, 54

C. Cachin, K. Kursawe, and V. Shoup. Random oracles in Constantinople: Practical asynchronous Byzantine agreement using cryptography. *J. Cryptology* 18(3):219–246, 2005. 47

R. Canetti. *Studies in Secure Multiparty Computation and Applications*. PhD thesis, Weizmann Institute of Science, 1996. 47

R. Canetti. Security and composition of multiparty cryptographic protocols. *J. Cryptology* 13(1):143–202, 2000. 35

R. Canetti. Universally composable security: A new paradigm for cryptographic protocols. In *42nd Annual Symposium on Foundations of Computer Science, FOCS 2001, Las Vegas, Nevada, USA, October 14–17, 2001*, pages 136–145. IEEE Computer Society, 2001. 29, 30, 32, 35, 38

R. Canetti, U. Feige, O. Goldreich, and M. Naor. Adaptively secure multi-party computation. In Miller [1996], pages 639–648. 48

R. Canetti, R. Pass, and A. Shelat. Cryptography from sunspots: How to use an imperfect reference string. In Sinclair [2007], pages 249–259. 39

R. Canetti and T. Rabin. Fast asynchronous Byzantine agreement with optimal resilience. In Kosaraju et al. [1993], pages 42–51. 47

M. Castro and B. Liskov. Practical Byzantine fault tolerance and proactive recovery. *ACM Trans. Comput. Syst.* 20(4):398–461, 2002. 47

D. Chaum. Untraceable electronic mail, return addresses, and digital pseudonyms. *Commun. ACM* 24(2):84–88, 1981. 37

D. Chaum, C. Crépeau, and I. Damgård. Multiparty unconditionally secure protocols (abstract) (informal contribution). In *STOC '88, Proceedings of the 28th Annual ACM Symposium on Theory of Computing*, page 462, 1987. 34

D. Chaum, C. Crépeau, and I. Damgård. Multiparty unconditionally secure protocols (extended abstract). In J. Simon, editor, *Proceedings of the 20th Annual ACM Symposium on Theory of Computing, May 2–4, 1988, Chicago, Illinois, USA*, pages 11–19. ACM, 1988. 38

J. Chen and S. Micali. Algorand: A secure and efficient distributed ledger. *Theor. Comput. Sci.* 777:155–183, 2019. 49, 52, 55, 57, 58

B. Chor and C. Dwork. Randomization in Byzantine agreement. *Advances in Computing Research* 5:443–497, 1989. 38

B. Chor, S. Goldwasser, S. Micali, and B. Awerbuch. Verifiable secret sharing and achieving simultaneity in the presence of faults (extended abstract). In *26th Annual Symposium on Foundations of Computer Science, Portland, Oregon, USA, 21–23 October 1985*, pages 383–395. IEEE Computer Society, 1985. 45

B. A. Coan and J. L. Welch. Modular construction of nearly optimal Byzantine agreement protocols. In P. Rudnicki, editor, *Proceedings of the 8th Annual ACM Symposium on Principles of Distributed Computing, Edmonton, Alberta, Canada, August 14–16, 1989*, pages 295–305. ACM, 1989. 46

R. Cohen, S. Coretti, J. A. Garay, and V. Zikas. Probabilistic termination and composability of cryptographic protocols. In M. Robshaw and J. Katz, editors, *Advances in Cryptology—CRYPTO 2016—36th Annual International Cryptology Conference, Santa Barbara, CA, USA, August 14–18, 2016, Proceedings, Part III*, volume 9816 of *LNCS*, pages 240–269. Springer, 2016. 45, 46, 55

J. Considine, M. Fitzi, M. Franklin, L. A. Levin, U. Maurer, and D. Metcalf. Byzantine agreement given partial broadcast. *J. Cryptol.* 18(3):191–217, July 2005. 50

S. Coretti, J. A. Garay, M. Hirt, and V. Zikas. Constant-round asynchronous multi-party computation based on one-way functions. In J. H. Cheon and T. Takagi, editors, *Advances in Cryptology—ASIACRYPT 2016—22nd International Conference on the Theory and Application of Cryptology and Information Security, Hanoi, Vietnam, December 4–8, 2016, Proceedings, Part II*, volume 10032 of *LNCS*, pages 998–1021, 2016. 38

B. David, P. Gazi, A. Kiayias, and A. Russell. Ouroboros Praos: An adaptively-secure, semi-synchronous proof-of-stake blockchain. In Nielsen and Rijmen [2018], pages 66–98. 57

F. Dold and C. Grothoff. Byzantine set-union consensus using efficient set reconciliation. *EURASIP Journal on Information Security* 2017(1):14, July 2017. 54

D. Dolev and R. Reischuk. Bounds on information exchange for Byzantine agreement. *J. ACM* 32(1):191–204, 1985. 46

D. Dolev, R. Reischuk, and H. R. Strong. Early stopping in Byzantine agreement. *J. ACM* 37(4):720–741, 1990. 44, 45

D. Dolev and H. R. Strong. Authenticated algorithms for Byzantine agreement. *SIAM J. Comput.* 12(4):656–666, 1983. 43, 44, 46, 49, 50

C. Dwork, N. A. Lynch, and L. J. Stockmeyer. Consensus in the presence of partial synchrony. *J. ACM* 35(2):288–323, 1988a. 38, 46, 47

C. Dwork and Y. Moses. Knowledge and common knowledge in a Byzantine environment: Crash failures. *Inf. Comput.* 88(2):156–186, 1990. 44

C. Dwork and M. Naor. Pricing via processing or combatting junk mail. In *Advances in Cryptology—CRYPTO 1992, 12th Annual International Cryptology Conference*, pages 139–147, 1992. 41

C. Dwork, D. Peleg, N. Pippenger, and E. Upfal. Fault tolerance in networks of bounded degree. *SIAM J. Comput.* 17(5):975–988, 1988b. 36

P. Feldman. *Optimal algorithms for Byzantine agreement*. PhD thesis, Massachusetts Institute of Technology, 1988. 47

P. Feldman and S. Micali. An optimal probabilistic protocol for synchronous Byzantine agreement. *SIAM J. Comput.* 26(4):873–933, 1997. 38, 45, 47, 48, 50

M. J. Fischer and N. A. Lynch. A lower bound for the time to assure interactive consistency. *Inf. Process. Lett.* 14(4):183–186, 1982. 44

M. J. Fischer, N. A. Lynch, and M. Merritt. Easy impossibility proofs for distributed consensus problems. *Distributed Computing* (1):26–39, 1986. 43

M. J. Fischer, N. A. Lynch, and M. Paterson. Impossibility of distributed consensus with one faulty process. *J. ACM* 32(2):374–382, 1985. 38, 48

M. Fitzi. *Generalized communication and security models in Byzantine agreement*. PhD thesis, ETH Zurich, Zürich, Switzerland, 2003. 36, 43, 56

M. Fitzi and J. A. Garay. Efficient player-optimal protocols for strong and differential consensus. In *Proceedings of the 22nd Annual ACM Symposium on Principles of Distributed Computing, PODC 2003, Boston, Massachusetts, USA, July 13-16, 2003*, pages 211–220. ACM, 2003. 44

M. Fitzi and M. Hirt. Optimally efficient multi-valued Byzantine agreement. In E. Ruppert and D. Malkhi, editors, *Proceedings of the 25th Annual ACM Symposium on Principles of Distributed Computing, PODC 2006, Denver, CO, USA, July 23–26, 2006*, pages 163–168. ACM, 2006. 46

C. Ganesh and A. Patra. Broadcast extensions with optimal communication and round complexity. In G. Giakkoupis, editor, *Proceedings of the 2016 ACM Symposium on Principles of Distributed Computing, PODC 2016, Chicago, IL, USA, July 25–28, 2016*, pages 371–380. ACM, 2016. 46

J. A. Garay, J. Katz, C. Koo, and R. Ostrovsky. Round complexity of authenticated broadcast with a dishonest majority. In Sinclair [2007], pages 658–668. 45

J. A. Garay, J. Katz, R. Kumaresan, and H. Zhou. Adaptively secure broadcast, revisited. In C. Gavoille and P. Fraigniaud, editors, *Proceedings of the 30th Annual ACM Symposium on Principles of Distributed Computing, PODC 2011, San Jose, CA, USA, June 6–8, 2011*, pages 179–186. ACM, 2011. 48

J. A. Garay, A. Kiayias, and N. Leonardos. The Bitcoin backbone protocol: Analysis and applications. *IACR Cryptology ePrint Archive* 2014:765, 2014. 49, 53, 59

J. A. Garay, A. Kiayias, and N. Leonardos. The Bitcoin backbone protocol: Analysis and applications. In E. Oswald and M. Fischlin, editors, *Advances in Cryptology—EUROCRYPT 2015—34th Annual International Conference on the Theory and Applications of Cryptographic Techniques, Sofia, Bulgaria, April 26–30, 2015, Proceedings, Part II*, volume 9057 of *LNCS*, pages 281–310. Springer, 2015. 41, 48, 49, 50, 51, 52, 53, 56, 58, 59

J. A. Garay, A. Kiayias, and N. Leonardos. The Bitcoin backbone protocol with chains of variable difficulty. In Katz and Shacham [2017], pages 291–323, 2017a. 57, 58, 59

J. A. Garay, A. Kiayias, N. Leonardos, and G. Panagiotakos. Bootstrapping the blockchain—directly. *IACR Cryptology ePrint Archive* 2016:991, 2016. 59

J. A. Garay, A. Kiayias, N. Leonardos, and G. Panagiotakos. Bootstrapping the blockchain, with applications to consensus and fast PKI setup. In M. Abdalla and R. Dahab, editors, *Public-Key Cryptography—PKC 2018—21st IACR International Conference on Practice and Theory of Public-Key Cryptography, Rio de Janeiro, Brazil, March 25–29, 2018, Proceedings, Part II*, volume 10770 of *LNCS*, pages 465–495. Springer, 2018. 49, 51, 52, 58, 59

J. A. Garay, A. Kiayias, and G. Panagiotakos. Proofs of work for blockchain protocols. *IACR Cryptology ePrint Archive* 2017:775, 2017b. 41

J. A. Garay and Y. Moses. Fully polynomial Byzantine agreement for $n > 3t$ processors in $t + 1$ rounds. *SIAM J. Comput.* 27(1):247–290, 1998. 44

J. A. Garay and K. J. Perry. A continuum of failure models for distributed computing. In *Distributed Algorithms, 6th International Workshop, WDAG '92, Haifa, Israel, November 2–4, 1992, Proceedings*, pages 153–165. Springer Verlag, 1992. 49

Y. Gilad, R. Hemo, S. Micali, G. Vlachos, and N. Zeldovich. Algorand: Scaling Byzantine agreements for cryptocurrencies. In *Proceedings of the 26th Symposium on Operating Systems Principles, Shanghai, China, October 28–31, 2017*, pages 51–68. ACM, 2017. 58

G. Golan-Gueta, I. Abraham, S. Grossman, D. Malkhi, B. Pinkas, M. K. Reiter, D. Seredinschi, O. Tamir, and A. Tomescu. SBFT: A scalable decentralized trust infrastructure for blockchains. *CoRR*, abs/1804.01626, 2018. 55

O. Goldreich. *The Foundations of Cryptography—Volume 1, Basic Techniques*. Cambridge University Press, 2001. 29, 35

O. Goldreich, S. Micali, and A. Wigderson. Proofs that yield nothing but their validity and a methodology of cryptographic protocol design (extended abstract). In *FOCS '86, 27th Annual Symposium on Foundations of Computer Science*, pages 174–187, 1986. 34, 35

O. Goldreich, S. Micali, and A. Wigderson. How to play any mental game or A completeness theorem for protocols with honest majority. In A. V. Aho, editor, *Proceedings of the 19th Annual ACM Symposium on Theory of Computing, 1987, New York, New York, USA*, pages 218–229. ACM, 1987. 38

S. Halevi, Y. Lindell, and B. Pinkas. Secure computation on the web: Computing without simultaneous interaction. In P. Rogaway, editor, *Advances in Cryptology—CRYPTO 2011—31st Annual Cryptology Conference, Santa Barbara, CA, USA, August 14–18, 2011, Proceedings*, volume 6841 of *LNCS*, pages 132–150. Springer, 2011. 49

M. Hirt and P. Raykov. Multi-valued Byzantine broadcast: The $t < n$ case. In P. Sarkar and T. Iwata, editors, *Advances in Cryptology—ASIACRYPT 2014—20th International Conference on the Theory and Application of Cryptology and Information Security, Kaoshiung, Taiwan, R.O.C., December 7–11, 2014, Proceedings, Part II*, volume 8874 of *LNCS*, pages 448–465. Springer, 2014. 46

M. Hirt and V. Zikas. Adaptively secure broadcast. In *Advances in Cryptology—EUROCRYPT 2010, 29th Annual International Conference on the Theory and Applications of Cryptographic Techniques, French Riviera, May 30–June 3, 2010, Proceedings*, pages 466–485. 2010. 48

J. Katz and C.-Y. Koo. On expected constant-round protocols for Byzantine agreement. *Journal of Computer and System Sciences* 75(2):91–112, 2009. 45

J. Katz and H. Shacham, editors. *Advances in Cryptology—CRYPTO 2017—37th Annual International Cryptology Conference, Santa Barbara, CA, USA, August 20–24, 2017, Proceedings, Part I*, volume 10401 of *LNCS*. Springer, 2017. 60, 64, 65

A. Kiayias, A. Russell, B. David, and R. Oliynykov. Ouroboros: A provably secure proof-of-stake blockchain protocol. In Katz and Shacham [2017], pages 357–388. 55, 57, 58

V. King and J. Saia. Byzantine agreement in expected polynomial time. *J. ACM* 63(2):13:1–13:21, 2016. 46

S. R. Kosaraju, D. S. Johnson, and A. Aggarwal, editors. *Proceedings of the 25th Annual ACM Symposium on Theory of Computing, May 16–18, 1993, San Diego, CA, USA*. ACM, 1993. 60, 62

K. Kursawe and V. Shoup. Optimistic asynchronous atomic broadcast. In L. Caires, G. F. Italiano, L. Monteiro, C. Palamidessi, and M. Yung, editors, *Automata, Languages and Programming, 32nd International Colloquium, ICALP 2005, Lisbon, Portugal, July 11–15, 2005, Proceedings*, volume 3580 of *LNCS*, pages 204–215. Springer, 2005. 47

L. Lamport, R. E. Shostak, and M. C. Pease. The Byzantine Generals Problem. *ACM Trans. Program. Lang. Syst.* 4(3):382–401, 1982. 27, 34, 35, 36, 42, 43

Y. Lindell, A. Lysyanskaya, and T. Rabin. On the composition of authenticated Byzantine agreement. *J. ACM* 53(6):881–917, 2006. 45

N. A. Lynch. *Distributed Algorithms*. Morgan Kaufmann Publishers Inc., San Francisco, CA, USA, 1996. 38

S. Micali. ALGORAND: The efficient and democratic ledger. *CoRR*, abs/1607.01341, 2016.

A. Miller and J. J. LaViola. Anonymous Byzantine consensus from moderately-hard puzzles: A model for Bitcoin. Tech Report CS-TR-14-01, University of Central Florida, April 2014. 49

A. Miller, Y. Xia, K. Croman, E. Shi, and D. Song. The honey badger of BFT protocols. In E. R. Weippl, S. Katzenbeisser, C. Kruegel, A. C. Myers, and S. Halevi, editors, *Proceedings of the 2016 ACM SIGSAC Conference on Computer and Communications Security, Vienna, Austria, October 24–28, 2016*, pages 31–42. ACM, 2016. 47, 55

G. L. Miller, editor. *Proceedings of the 28th Annual ACM Symposium on the Theory of Computing, Philadelphia, Pennsylvania, USA, May 22–24, 1996*. ACM, 1996. 60, 61

S. Nakamoto. Bitcoin: A peer-to-peer electronic cash system. http://bitcoin.org/bitcoin.pdf, 2008a. 27

S. Nakamoto. The proof-of-work chain is a solution to the Byzantine Generals' problem. The Cryptography Mailing List, https://www.mail-archive.com/cryptography@metzdowd.com/msg09997.html, November 2008b. 49

S. Nakamoto. Bitcoin open source implementation of P2P currency. http://p2pfoundation.ning.com/forum/topics/bitcoin-open-source, February 2009. 27

M. Naor and M. Yung. Universal one-way hash functions and their cryptographic applications. In D. S. Johnson, editor, *Proceedings of the 21st Annual ACM Symposium on Theory of Computing Seattle, Washington, May 14–17, 1989, USA*, pages 33–43. ACM, 1989. 41

G. Neiger. Distributed consensus revisited. *Inf. Process. Lett.* 49(4):195–201, 1994. 35, 44

J. B. Nielsen and V. Rijmen, editors. *Advances in Cryptology—EUROCRYPT 2018—37th Annual International Conference on the Theory and Applications of Cryptographic Techniques, Tel Aviv, Israel, April 29–May 3, 2018 Proceedings, Part II*, volume 10821 of *LNCS*. Springer, 2018. 62, 66

M. Okun. Agreement among unacquainted Byzantine generals. In P. Fraigniaud, editor, *DISC*, volume 3724 of *LNCS*, pages 499–500. Springer, 2005a. 37, 48

M. Okun. Distributed computing among unacquainted processors in the presence of Byzantine failures. PhD thesis, Hebrew University of Jerusalem, 2005b. 48

R. Pass, L. Seeman, and A. Shelat. Analysis of the blockchain protocol in asynchronous networks. In *Advances in Cryptology—EUROCRYPT 2017—36th Annual International Conference on the Theory and Applications of Cryptographic Techniques, Paris, France, April 30–May 4, 2017, Proceedings, Part II*, pages 643–673. Springer, 2017. 53, 57, 58, 59

R. Pass and E. Shi. The sleepy model of consensus. In T. Takagi and T. Peyrin, editors, *Advances in Cryptology—ASIACRYPT 2017—23rd International Conference on the Theory and Applications of Cryptology and Information Security, Hong Kong, China, December 3–7, 2017, Proceedings, Part II*, volume 10625 of *LNCS*, pages 380–409. Springer, 2017. 51, 56, 57, 58

R. Pass and E. Shi. Thunderella: Blockchains with optimistic instant confirmation. In Nielsen and Rijmen [2018], pages 3–33. 58

A. Patra. Error-free multi-valued broadcast and Byzantine agreement with optimal communication complexity. In A. F. Anta, G. Lipari, and M. Roy, editors, *Principles of Distributed Systems—15th*

International Conference, OPODIS 2011, Toulouse, France, December 13–16, 2011, Proceedings, volume 7109 of *LNCS*, pages 34–49. Springer, 2011. 46

A. Patra, A. Choudhury, and C. P. Rangan. Asynchronous Byzantine agreement with optimal resilience. *Distributed Computing* 27(2):111–146, 2014. 47

M. C. Pease, R. E. Shostak, and L. Lamport. Reaching agreement in the presence of faults. *J. ACM* 27(2):228–234, 1980. 27, 34, 54

B. Pfitzmann and M. Waidner. Unconditional Byzantine agreement for any number of faulty processors. In *STACS 92, 9th Annual Symposium on Theoretical Aspects of Computer Science Proceedings*, volume 577, pages 339–350. Springer, 1992. 40, 46

M. O. Rabin. Randomized Byzantine Generals. In *FOCS 24th Symposium on Foundations of Computer Science (1983), Tucson, AZ, USA, November 7–9, 1983*, pages 403–409. IEEE Computer Society, 1983. 38, 40, 45

F. B. Schneider. Implementing fault-tolerant services using the state machine approach: A tutorial. *ACM Comput. Surv.* 22(4):299–319, Dec. 1990. 28, 54

A. Sinclair, editor. *48th Annual IEEE Symposium on Foundations of Computer Science (FOCS 2007), Providence, RI, USA, October 20–23, 2007, Proceedings*. IEEE Computer Society, 2007. 62, 64

N. Stifter, A. Judmayer, P. Schindler, A. Zamyatin, and E. R. Weippl. Agreement with Satoshi—on the formalization of Nakamoto consensus. *IACR Cryptology ePrint Archive* 2018:400, 2018. 28

R. Turpin and B. A. Coan. Extending binary Byzantine agreement to multivalued Byzantine agreement. *Inf. Process. Lett.* 18(2):73–76, 1984. 34

E. Upfal. Tolerating linear number of faults in networks of bounded degree. In N. C. Hutchinson, editor, *Proceedings of the 11th Annual ACM Symposium on Principles of Distributed Computing, Vancouver, British Columbia, Canada, August 10–12, 1992*, Pages 83–89. ACM, 1992. 36

A. C.-C. Yao. Protocols for secure computations. *Proceedings of the 23rd Annual Symposium on Foundations of Computer Science (SFCS 1982)*, pages 160–164, 1982. DOI: 10.1109/SFCS.1982.38. 34

AUTHORS' BIOGRAPHIES

Juan A. Garay is a full professor at Texas A&M University's Computer Science and Engineering Department. Previously, after receiving his PhD in Computer Science from Penn State, he was a postdoc at the Weizmann Institute of Science (Israel), and held research positions at the IBM T.J. Watson Research Center, Bell Labs, AT&T Labs–Research, and Yahoo Research. His research interests include both foundational and applied aspects of cryptography and information security; in more detail, they include cryptographic protocols and schemes, secure multi-party computation, blockchain protocols and cryptocurrencies, cryptographic and game theory, and consensus problems. He is the author of over 170 published works (including articles, patents, and edited volumes) in the areas of cryptography, network security, distributed computing, and algorithms; has been involved in the design, analysis, and implementation of a variety of secure systems; and is the recipient of a Thomas A. Edison Patent Award, two Bell Labs Teamwork Awards, an IBM Outstanding Technical Achievement Award, and an IBM Research Division Award. Dr. Garay has served on the program committees of numerous conferences and international panels—including co-chairing Crypto 2013 and 2014, the discipline's premier conference. He is a Fellow of the International Association for Cryptologic Research (IACR).

Aggelos Kiayias is chair in Cyber Security and Privacy and director of the Blockchain Technology Laboratory at the University of Edinburgh. He is also the Chief Scientist at blockchain technology company IOHK. His research interests are in computer security, information security, applied cryptography, and foundations of cryptography with a particular emphasis in blockchain technologies and distributed systems, e-voting, and secure multi-party protocols, as well as privacy and identity management. His research has been funded by the Horizon 2020 programme (EU), the European Research Council (EU), the Engineering and Physical Sciences Research Council (UK), the Secretariat of Research and Technology (Greece), the National Science Foundation (USA), the Department of Homeland Security (USA), and the National Institute of Standards and Technology (USA). He has received an ERC Starting Grant, a Marie Curie fellowship, an NSF Career Award, and a Fulbright Fellowship. He holds a PhD from the City University of New York and is a graduate of the Mathematics department of the University of Athens. He has over 100 publications in journals and conference proceedings in the area of cryptography. He has served as the program chair of the Cryptographers' Track of the RSA conference in 2011 and the Financial Cryptography and Data Security conference in 2017, as well as the general chair of Eurocrypt 2013. He also served as the program chair of Real World Crypto Symposium 2020 and the Public-Key Cryptography Conference 2020.

CHAPTER 3

The Next 700 Smart Contract Languages

Ilya Sergey, *Yale-NUS College and National University of Singapore, Singapore*

3.1 INTRODUCTION

Smart contracts are a mechanism for expressing replicated computations powered by a decentralized consensus protocol [Szabo 1994]. They are most commonly used to define custom logic for transactions operating over a blockchain—that is, a decentralized Byzantine-fault-tolerant distributed ledger [Bano et al. 2019, Pîrlea and Sergey 2018]. In addition to a typical state of computations, a blockchain stores a mapping from *accounts* (public keys or addresses) to quantities of *tokens* owned by said accounts. Execution of an arbitrary program (aka a smart contract) is done by *miners*, who run the computations and maintain the distributed ledger in exchange for a combination of *gas* (transaction fees based on the execution length, denominated in the intrinsic tokens and paid by the account calling the smart contract) and *block rewards* (inflationary issuance of fresh tokens by the underlying protocol). One distinguishing property of smart contracts, not found in standard computational settings, is the management of token transfers between accounts. While simple forms of smart contracts were already available for regulating exchange of virtual coins in earlier cryptocurrencies such as Bitcoin [Nakamoto 2008], smart contracts owe their wide adoption to the Ethereum framework [Buterin 2013, Wood 2014]. Since their first public implementations in the mid-2010s, protocols supporting smart contract deployment have found many applications in digital finance, accounting, voting, gaming, and many other areas that naturally require decentralization.

One of the challenging design aspects of smart contracts is the fact that the outcomes and their executions are determined by their interactions with a decentralized adversarial environment. That is, once deployed, smart contracts might exchange data with other contracts, possibly designed to exploit vulnerabilities in their logic, with the aim to provoke an unforeseen token transfer or to execute a denial-of-service attack [Luu et al. 2016a]. Such vulnerabilities are typically caused by rather subtle contract behavior that diverges from the "intuitive understanding" of the language in the minds of the contract developers. Some of the most prominent attacks on smart contracts deployed on Ethereum blockchain—for example, the attack on the DAO[1] [del Castillo 2016] and Parity wallet [Alois 2017] contracts—have exploited some undocumented or poorly understood behavior of certain language features. These and many other reported attacks on the Ethereum-deployed contracts [Atzei et al. 2017, Kalra et al. 2018, Kolluri et al. 2019, Krupp and Rossow 2018, Nikolić et al. 2018, Rodler et al. 2019] have made *execution safety and formal correctness guarantees* into the primary concerns for smart contract–based programming.

The fact that third-party smart contracts cannot be trusted has another important implication for the design of the implementation language semantics. The decentralized nature of contract execution means that the majority of the involved miners will have to agree on the outcome of a transaction involving a contract's invocation. Therefore, by providing the runtime support for contract executions, the miners become open to a wide class of denial-of-service attacks. Such attacks may be effectively implemented by contracts that never terminate under certain conditions or use an extensive amount of memory. As a remedy to this challenge, in Ethereum's pioneering approach to smart contracts every transaction costs a certain amount of *gas* [Buterin 2013, Wood 2014], a monetary value in Ethereum's currency, paid by a transaction-proposing party. Computations (performed by invoking smart contracts) that require more computational or storage resources cost more gas than those that require fewer resources. Gas cost is deduced dynamically: each execution step is being charged from the gas supply paid for; if a transaction "runs out of gas" in the midst of its execution, it is interrupted, with all the corresponding changes discarded. Therefore, an adequate model for principled gas accounting is crucial for the definition of smart contract language semantics and reasoning about the safety and the dynamic costs of its executions. As the recent research shows, miscalculated gas costs, both at the level of the language and of the individual contracts, are quite common [Albert et al. 2019, Chen et al. 2017, Grech et al. 2018, Marescotti et al. 2018, Pérez and Livshits 2019], and they might lead to even more severe vulnerabilities to exploit.

At the beginning of smart contract adoption, it was hard to predict what kinds of applications clients would most likely be willing to develop and deploy. While the encodings of financial routines form the majority of smart contracts to date, Ethereum's design provided a very expressive runtime environment for smart contracts—the Ethereum Virtual Machine (EVM) [Wood 2014]. EVM offered a Turing-complete low-level language whose features, amongst others, included arbitrary interaction between contracts (with any contract's code accessible to any other contract), dynamic contract creation, and ability to use introspection on the entire state of the Ethereum blockchain. Such versatility made EVM a very popular platform for developing high-level languages to compile to, which in turn resulted in an explosion of Ethereum applications ranging from fully decentralized auctions, fundraisers, and crowdfunding to multiplayer games, protocols for verified computations,

---

[1] Decentralized Autonomous Organization [Ethereum Foundation 2018a].

and even fraud schemes [Dong et al. 2017, McCorry et al. 2017, McCorry et al. 2019]. The *expressivity* and a low-level language design are a double-edged sword. While offering great flexibility in implementing custom transaction logic, this environment is frequently at odds with the safety aspect outlined below. For instance, Ethereum contracts, deployed in a low-level language, render independent audit and formal verification of deployed code infeasible in practice. Narrowing the scope of the minimal necessary functionality that smart contracts need to possess is thus an active subject for ongoing research.

The history of programming language (PL) design and implementation is also a history of making programs run *faster*, by exploiting the corresponding problem domains as well as underlying hardware architectures. In this regard, smart contracts are seemingly no different from other programs, and optimizing their runtime will also benefit the entire protocol. However, in light of the aforementioned safety and expressivity aspects, optimizations of smart contracts pose a number of unusual challenges for the language and runtime designers. For instance, it is not clear how optimizations will interact with the gas costs defined for a particular execution mode. Furthermore, to a large extent the smart contract runtimes to date treat the underlying consensus protocols as a black box, without taking any advantage of their architecture, which might allow for parallel execution of transactions [Luu et al. 2016b, Kokoris-Kogias et al. 2018, Al-Bassam et al. 2018].

3.1.1   WHAT WE WILL DISCUSS

The three dimensions of smart contract language design can thus be summarized by a diagram shown in Figure 3.1. As a reference point, we illustrate Ethereum's EVM design choices by a dashed line, emphasizing its focus on expressivity and optimization-friendliness, but not so much on formal guarantees of execution safety.

In this chapter we aim to provide tentative answers to the following questions:

1. What are the essential concepts of smart contracts that by all means need to be represented in the language used to implement them?

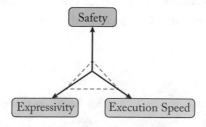

FIGURE 3.1: Language design trade-off. The dashed line shows EVM's design choices.

2. Which of the well-established PL techniques can be useful for this task and what are the challenges in adapting them?

3. What are the unsolved problems in smart contract language design that one should consider tackling in the future?

In the rest of this chapter, we will elaborate on these three dimensions, discussing various components specific to smart contract programming, and outlining multiple possibilities for programming language abstractions targeting greater expressivity, safety, or providing a more suitable ground for program optimizations.

3.1.2   WHAT WE WILL NOT DISCUSS

At the time of this writing, the smart contract programming landscape is growing at a breakneck speed, with new language proposals emerging nearly every week. To date, most of those languages are available in the form of a sparsely documented repository, a position paper, or a blog post [æternity Blockchain 2019, Alfour 2019, Coblenz 2017, Hirai 2018, IOHK Foundation 2019a, RChain Cooperative 2019, Reitwiessner 2017]. Therefore, we do not aim to provide a detailed survey of the currently available smart contract programming technology, but rather focus on the conceptual components that either might have been or have been encompassed in some of those proposals (in which case appropriate examples will be provided).

In the past few years, the efforts toward discovery, analysis, modeling, and fixing specific classes of vulnerabilities in Ethereum smart contracts have turned into an active research field [Alt and Reitwießner 2018, Amani 2018, Bansal et al. 2018, Bhargavan et al. 2016, Chang et al. 2018, Grech et al. 2018, Grishchenko et al. 2018, Grossman et al. 2018, Kalra et al. 2018, Kolluri et al. 2019, Krupp and Rossow 2018, Luu et al. 2016a, Marescotti et al. 2018, Nikolić et al. 2018, Tikhomirov et al. 2018, Tsankov et al. 2018]. While some of those techniques are informative for PL design for smart contracts, their formulation is, in most cases, very specific to the Ethereum platform and EVM. The survey of those approaches and tools is, thus, beyond the scope of this chapter as well. Readers interested in the contemporary state of the art in those directions are encouraged to check the corresponding survey papers [Angelo and Salzer 2019, Seijas et al. 2016].

3.2   BACKGROUND

To set the stage for the discussion on smart contract language design, let us first consider a simple smart contract and understand its behavior and properties. Figure 3.2 shows one of the most common applications of smart contracts—a crowdfunding campaign—implemented in Ethereum's

```
contract Crowdfunding {
  address owner;
  uint256 deadline;
  uint256 goal;
  mapping(address => uint256) backers;

  function Crowdfunding(uint256 numberOfDays, uint256 _goal) public {
    owner = msg.sender;
    deadline = now + (numberOfDays * 1 days);
    goal = _goal;
  }

  function donate() public payable {
    require(now < deadline);         // before the fundraising deadline

    backers[msg.sender] += msg.value;
  }

  function claimFunds() public {
    require(address(this).balance >= goal); // funding goal met
    require(now >= deadline);                // after the withdrawal period
    require(msg.sender == owner);

    msg.sender.call.value(address(this).balance)();
  }

  function getRefund() public {
    require(address(this).balance < goal);  // campaign failed: goal not met
    require(now >= deadline);                // in the withdrawal period

    uint256 donation = backers[msg.sender];
    backers[msg.sender] = 0;
    msg.sender.call.value(donation)();
  }
}
```

FIGURE 3.2: A crowdfunding contract in Solidity.

Solidity programming language [Ethereum Foundation 2019].[2] Solidity is a high-level language with syntax similar to JavaScript, and it compiles directly to EVM. As of early 2020, this is a de facto programming language for smart contracts that has received wide adoption due to the popularity of the Ethereum protocol.

The contract in Figure 3.2 is very similar to a stateful object in a language such as Java or C#. It features four mutable *fields*. The field `owner` of type `address` defines the identity of the account that deploys the contract. The fields `goal` and `deadline` set the main parameters of the crowdfunding campaign: the amount of currency it aims to raise and the deadline (i.e., the final block) after which donations are no longer accepted. Finally, the field `backers` of type `mapping (address => uint256)` is a mutable hash-map that stores the amounts donated by different backers identified by their account addresses.

The *constructor* `Crowdfunding` of the contract sets the fields `owner`, `deadline`, and `goal` to the values provided upon the contract deployment. The value of the `owner` is retrieved from the field `sender` of the implicit constructor argument `msg` that denotes the message initiating the interaction with the contract (in this case, its deployment) in the corresponding transaction. Thus, `msg.sender` refers to the account that has initiated the transaction. Upon the successful execution of the constructor, the resultant state and the code of the contract are replicated amongst the miner nodes.

All the subsequent interactions with the contract by the third parties are done by means of invoking its `functions` (methods), of which it has three. The first one, `donate`, allows transfer of the donation to the contract. The first line of the method checks, via the `require` construct, that the current block (referred to as `now`) is strictly smaller than the `deadline`. If this test fails, the whole transaction is reverted. The amount of currency transferred to the contract from the party initiating the interaction is implicitly stored in the attribute `value` of the incoming message. As the function is marked `payable`, this amount will be *implicitly* added to the contract's balance—the code does not contain instructions to do so. For the purpose of correct accounting, the method records to the map `backers` the donated value attributed to the message sender.

The purpose of the method `claimFunds` is to allow the owner of the contract to transfer all the funds from the contract to its own account. The method first performs a number of checks—ensuring that the collected balance of the contract is larger than or equal to the set `goal`, and that the deadline has passed, as well as that the sender of the message is indeed the initial `owner`—and then the contract performs the transfer of funds to the sender of the message (i.e., the owner) via a somewhat unusual Solidity construct `call.value(...)()`. On the other hand, if when conditions

---

[2] This contract is adapted from https://programtheblockchain.com/posts/2018/01/19/writing-a-crowdfunding-contract-a-la-kickstarter/.

are checked the deadline has passed but the goal has not been reached, the final method, getRefund, allows the backers to retrieve their donations, transferring the correct amounts of currencies back to their own accounts. If this is indeed the case, the donation amount is retrieved from the backers map, and then transferred to the backer in the last line of the method.

Altogether the three methods of the Crowdfunding contract form its interface, defining all modes of interaction with it. It is important to notice that, by nature, all smart contracts are *passive*— that is, they do not engage in any interactions pro-actively. Instead, the code in their methods is executed in reaction to the messages sent by the external accounts, which might belong to users or to other contracts serving as interaction "proxies."

3.2.1   THE SUBTLETIES OF THE CROWDFUNDING CONTRACT

While seemingly very simple, the Crowdfunding contract has a number of intricacies, and misunderstanding of any of those might lead to the deployment of a flawed implementation, which will result in a potential loss of funds.

Consider, for instance, the very first donate method. Its only purpose is to check that the current block has a number smaller than the set deadline, and, if that is the case, accept the incoming funds from the backer. The subtle point in this logic is that the map backers might already store a previous donation, and, hence, the new donation needs to be added to it. This logic is implemented correctly in line 16 via the += operator. That being said, by a very trivial oversight, one could instead write backers[msg.sender] = donation. This logical mistake would not be caught by the compiler, yet it would result in the backer losing its previous donation, which would be irretrievable. How so? Consider the logic of the getRefund method. The donation amount transferred back to the backer is taken from the backers map, which, given the mistake described above, would only store the *latest* donation by the backer! That is, the amount donated previously would no longer be accounted for and would not be returned to the backer. Quite ironically, the described programming mistake would not prevent the owner from cashing out the funding collected in the case of a successful campaign. This is because the code in line 24 of Figure 3.2 operates with the contract's balance rather than the contents of the backers map, thus depleting the entire balance accumulated by the contract through previously accepted donations.

Another possible bug that would render the Crowdfunding contract useless would be if the programmer forgot to put the instruction backers[msg.sender] = 0 in line 32. In this case, the backer would be able to get the refund several times, possibly until the contract's balance was depleted.

Swapping lines 32 and 33 would lead to even more interesting behavior. According to Solidity's semantics, the execution of the command msg.sender.call.value(donation)() would transfer not only the funds but also the control over execution to the account msg.sender, thus allowing it to perform some additional operations before returning the control to the Crowdfunding contract. Specifically, the account of `msg.sender` could belong to another contract, which could invoke `getRefund` again, thus withdrawing more funds, similarly to the previous scenario in which we forgot to set the backer's donation to 0. This behavior is known as *re-entrancy vulnerability* [Gün Sirer 2016]. It has been a source of the most famous bug in a contract deployed in Ethereum [del Castillo 2016], sparking a line of research targeting prevention of these mistakes in the future [Grossman et al. 2018, Kalra et al. 2018, Tsankov et al. 2018].[3]

3.2.2   REASONING ABOUT CONTRACT PROPERTIES

Smart contracts are safety-critical applications. While existing static analysis tools help to significantly reduce the risk of deploying a faulty contract [Kalra et al. 2018, Permenev et al. 2020, Securify 2019], when designed for a language as complex as Solidity/EVM they will inevitably be unsound (i.e., they might miss some bugs). By designing a language that avoids certain mistakes in programs by construction, one can achieve stronger correctness guarantees.

As demonstrated by the `Crowdfunding` example, despite their simple logic of state manipulation, smart contracts are not always easy to get right. Not only must the programmer keep in mind all the properties relevant to the normal contract behavior prior to its deployment, but she also needs to be aware of the precise semantics of certain language constructs in order to avoid unexpected outcomes. This complexity is exacerbated by the fact that, once deployed (i.e., replicated via the blockchain), smart contracts cannot be patched or amended.

One way to ensure that the contract obeys some "common sense" before it is deployed is to state its properties and ensure that the code preserves them. Those properties are commonly phrased as contract *invariants*—that is, the assertions that hold at any point of the contract's lifetime. For instance, the following set of formal properties comprise a reasonably complete specification of the `Crowdfunding` contract:

P1.   (No leaking funds) The contract's accounted funds do not decrease unless the campaign has been funded or the deadline has expired.

P2.   (Donation record preservation) The contract preserves cumulative records of individual donations by backers, unless they interact with it.

P3.   (The backer can get refunded) If the campaign fails, the backers can eventually get their refund for the whole amount they have donated. They can get this refund exactly once.

Recent works have shown how to formally *prove* these properties using machine-assisted tools [Coq Development Team 2019] for a small smart contract language [Sergey et al. 2018a,

---

[3] As the result of this bug, Solidity has been augmented with a number of restrictive primitives for transferring funds, namely `send` and `transfer`, which are preferable in most similar scenarios.

2018b]. However, those ironclad correctness guarantees are only available if the *entire language*'s semantics is well-defined and formalized. This semantics should describe, among other execution artefacts, contract deployment and interaction with other entities in the blockchain. As of now, no fully formal semantics of Solidity exists, and this is why existing verification tools [Permenev et al. 2020] for Ethereum have to rely on the ad hoc understanding of the language's runtime.

3.2.3   CONTRACT EXECUTION MODELS

Even though contracts are to a large extent just stateful replicated objects, the choice of the language paradigm for implementing them should not be solely based on this fact. While JavaScript-like syntax and semantics have been chosen for Solidity to simplify its adoption, the same choice makes formal reasoning and efficient compilation far from straightforward. For instance, it has been observed that most of the contracts' functionality in fact falls into the pure functional fragment of data manipulation, with state manipulation comprising only a small part of the contract implementations. This informed the design of several contract languages, such as Michelson [Tezos Foundation 2018], Liquidity [OCaml PRO 2019], and Scilla [Sergey et al. 2019].

Another factor determining the choice of programming abstractions is the *state model* that is supported by the underlying blockchain consensus protocol. For instance, Ethereum follows an *Account/Balance* model, wherein the state of the user and the contracts (including funds attributed to them) is stored in a database-like fashion, with account addresses serving as unique keys. An alternative model, adopted by Bitcoin [Nakamoto 2008] and Cardano [Kiayias et al. 2017] blockchains, is based on the *unspent transaction outputs* (UTXO) model [Sun 2018], wherein transactions form a directed acyclic graph, threading the state of individual accounts from the initial until the terminal nodes. The UTXO model has been shown to be a better fit for the functional programming model [Chakravarty et al. 2020].

3.2.4   GAS ACCOUNTING

In the Crowdfunding example from Figure 3.2, all methods of the contract are straight-line code, with no loops or recursion. However, those features are frequently necessary for implementing common machinery. For instance, executing an iteration over a list of backers' account addresses would allow for reimbursing all of them in a single call instead of calling getRefund for each backer individually.

Unfortunately, statically unbounded iteration, as well as general loops and recursion, would be a source of denial-of-service attacks on the entire blockchain protocol, as all miners would have to run code that potentially cannot terminate. As of now, the community has reached a consensus that precisely defines execution cost semantics, aka *gas*, to be an inherent part of any smart contract language to be used in an open system, where arbitrary parties can join. That said, allocating adequate gas costs is far from straightforward, and to date this aspect of contract language design has received little attention from the research community. In Section 3.6, we will discuss some non-obvious challenges when assigning gas costs to smart contract executions, as well as programming language–enabled techniques for analyzing gas usage patterns.

3.2.5 ON THE ROLE OF TYPES

Types in programming languages are going to be a paramount motif of our overview [Pierce 2002]. Modern type systems provide a compositional syntactic approach for ensuring strong syntactic guarantees for a variety of execution properties. We will see how type-based approaches help the contract developers to ensure stronger atomicity guarantees (Section 3.3), restrict communication patterns (Section 3.4), and enforce the correct handling of digital assets (Section 3.5), and reason about resource consumption (Section 3.6).

3.3 ENFORCING CONTRACT INVARIANTS

The re-entrancy vulnerability in the DAO contract [del Castillo 2016, Gün Sirer 2016] is a great example of violating one of the foremost principles in designing software that might interact with other untrusted components—*preservation of invariants* [Sergey and Hobor 2017]. Invariants are logical assertions that postulate certain relations between the components of the contract. For instance, the property P2 of the `Crowdfunding` contract is an invariant that can be expressed as the following mathematical assertion:

$$\texttt{now} \geq \texttt{deadline} \vee \sum_{b \in \mathrm{dom(backers)}} \texttt{backers(b)} = \texttt{balance} \tag{3.1}$$

Invariants are a common way to reason about the validity of a state of an evolving object at any point in its lifetime, and invariant-based reasoning is customarily used to argue for the correctness of mutable concurrent data structures [Herlihy and Shavit 2008]. In such objects, each modification made to the object's state by a single process should be performed *atomically*—that is, so that it cannot be interrupted by other processes in the midst of its execution. A concurrent object's methods are implemented in such a way that they may only violate the object's invariants during their execution, but must restore them at the end of the call. Consider Figure 3.3. Its top part shows an interaction between a concurrent object c and its *environment*. Whenever the method c.atomicMethod() is executed, it assumes a certain invariant, and, upon termination, it restores it. This way, the next time the same method is called, it may still rely on the invariant being preserved. Taking another look at the `Crowdfunding` contract in Figure 3.2, we can see that the invariant (3.1) indeed holds after the contract is initialized by its constructor, and all methods preserve it. Thus, one way to look at the contract is as a valid *concurrent object*, which maintains its own state invariants.

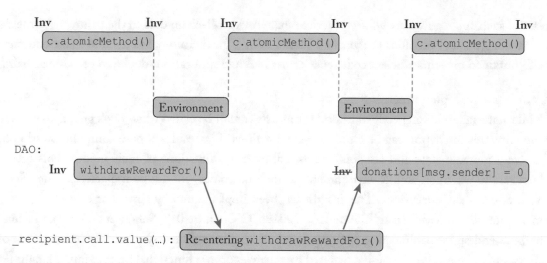

FIGURE 3.3: Invariant-preserving interactions with an atomic object (top). Invariant-violating re-entrant behavior in the DAO contract (bottom).

This principle is, however, violated by the DAO contract, whose re-entrant execution is schematically shown in the bottom of Figure 3.3. In the middle of the call to the `withdraw-RewardFor()` method, a method of another contract is called via `_recipient.call.value(...)`, and, upon returning the execution back to `withdrawRewardFor()`, the invariant is not restored. This lack of atomicity made it possible for the adversaries to exploit the "dirty" state of the contract and deplete it of its funds by calling the `withdrawRewardFor()` method again, in a re-entrant way [Sergey and Hobor 2017].

Issues of this nature could be avoided should the language design be more restrictive than what's allowed in Solidity. For instance, in the Scilla language for smart contracts [Sergey et al. 2019], any *non-atomic* interactions between contracts are forbidden by design. That is, the only way for a contract to call another contract is to first finish its own execution and only then pass the control to the other. While this design does not guarantee that the invariants will be preserved (as those tend to be complex and depend on the domain), it eliminates the DAO-like re-entrancy scenario entirely.

The downside of by-construction atomicity

The language-enforced atomicity comes with a cost, as it requires the developer to design a contract in a way that makes all its modifications self-contained and not dependent on the intermediate interaction with other contracts. This shortcoming does not exclude the possible implementation of most typical contract applications, but makes some of them quite cumbersome. A particular class of such applications are *oracles*—services that provide off-chain data to the contracts by means of invoking *callback* methods [Sergey and Hobor 2017]. Furthermore, the need to make contracts self-contained prevents efficient code reuse across multiple applications deployed on a blockchain.

Code reuse with pure functions

While state-manipulating code of external contracts cannot be called when one's goal is to enforce atomicity, this restriction can be lifted for *pure* functions. The results of pure computations do not involve a mutable state and are obtained as mathematical functions of their inputs. This makes them safe to use in an atomic environment, as their outcomes will not be impacted by the effects of other contracts' executions. This insight has been implemented by a number of contemporary smart contract languages [IOHK Foundation 2019b, O'Connor 2017, Sergey et al. 2019], in which the pure and state-manipulating computations can be distinguished syntactically or by means of an expressive type system (a design inspired by general-purpose functional programming languages such as Standard ML and Haskell).

Type systems for invariant preservation

General coherence properties of smart contracts can often be captured and enforced by means of type systems. Sometimes the contract invariants can be so generic that they might be captured in a form of types. For instance, the post-deadline configuration of the Crowdfunding contract can be described as a form of *typestate* [Aldrich et al. 2009]—an approach allowing the inclusion of state information in types of variables to indicate which can be modified (or not) while the object is in that state. For instance, one can define a typestate PostDeadline as the only one in which it is allowed for the contract to send money to third parties. The typestate-based approach has been implemented in the Flint [Schrans 2018, Schrans et al. 2018] and Obsidian [Coblenz et al. 2019] languages to provide stronger guarantees for contracts implemented in the Solidity style object-based model.

3.4   STRUCTURING COMMUNICATION

The need to enforce atomicity of contracts also hints at a solution for arranging interactions between them.

Solidity has leaned toward the object-based model, familiar to seasoned Java and C# developers, in which all contract interactions are simply method calls. As we've seen above, this model makes it non-trivial to enforce by-construction atomicity and requires additional mental efforts to structure the contract in a way that would always preserve its invariants.

A more suitable approach for this purpose is to implement interaction protocols between contracts as *communication* via passing messages. The message-passing paradigm for building applications of multiple interacting entities is well studied from both the theoretical and practical

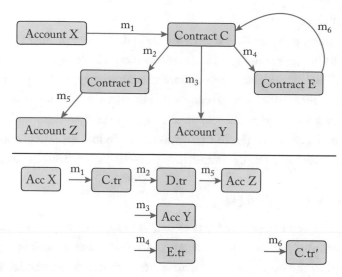

FIGURE 3.4: An interaction between accounts and contracts within a transaction (top), and its sequentialization when executed by the protocol (bottom).

perspectives. The most notable theoretical frameworks that describe message-passing programs are π-calculus [Milner 1980], actors [Agha 1990], and input/output automata [Lynch and Tuttle 1989]. Practical implementation of those concepts can be found in general-purpose programming languages, such as Erlang [Armstrong 2007] and Scala [Haller and Sommers 2011]. The idea of implementing smart contracts as communicating state-transition systems was first explored in the SCILLA [Sergey et al. 2018a] language developed for ZILLIQA blockchain [Zilliqa Team 2017]. A contract in SCILLA is a definition of mutable and immutable state components (with the former defined upon deployment), as well as a number of "transitions," each serving as a handler reacting to a certain kind of messages. Transitions are atomic, and may result in sending more messages to other contracts, which will be processed later.

A transition invocation may trigger a chain of contract calls as shown in Figure 3.4 (top). In the case of a multi-contract transaction (i.e., when a contract interacts with other contracts), the emitted messages are sequentialized by following a *breadth-first* traversal of the transaction communication graph (bottom of figure). The messages are then executed in sequence.[4] The combined output of the set of messages resulting from a transaction is committed to the blockchain atomically, in the sense that nothing is committed unless all messages succeed. If one message completes and the next one runs out of gas, the entire transaction is rolled back.

---

[4] Breadth-first was chosen over depth-first as it provides better fairness guarantees for message processing.

Type systems for message passing

The idea of implementing contract interactions via explicit message passing and atomic transitions has been implemented in other languages for smart contracts: Rholang [RChain Cooperative 2019] and Nomos [Das et al. 2019]. Both languages feature an expressive type system that statically enforces certain user-provided interactions between contracts. For instance, Nomos' type system, inspired by resource-aware binary session types [Das et al. 2018], ensures that communication between two contracts spanning multiple messages and transitions will follow a particular protocol while also consuming a specified fixed amount of computational resources (cf. Section 3.6.1).

3.5   IT'S ALL ABOUT MONEY

After all, the main purpose of smart contracts to date is to manage digital assets. Solidity's take on considering money as simply an unsigned integer datatype is prone to make programmers commit errors that would lead to the loss of significant amounts of funds. A substantial amount of research effort has been dedicated to remedying this design choice, utilizing tools for automated analyses of Ethereum contracts. Such tools can check that, for instance, a contract does not transfer its own funds to any third party unconditionally (the so-called *prodigal* scenario) [Nikolić et al. 2018]. In this section we survey programming language techniques used to ensure the correct handling of digital money by a contract.

Cash-flow analysis

The SCILLA smart contract language has introduced a static analysis whose purpose is to ensure that the values representing money are never mixed with other values in an inconsistent way [Johannsen and Kumar 2019]. Specifically, the cash-flow analysis attempts to determine which parts of the contract's state (i.e., its fields) represent an amount of money, in order to ensure that money is being accounted for in a consistent way. To do this, the analysis employs standard techniques of abstract interpretation [Cousot and Cousot 1977], so each field, parameter, local variable, and subexpression in the contract is given a *tag* indicating if and how it is used w.r.t. representing money.

Running the analysis on the Crowdfunding contract (Figure 3.2) results in the fields goal and backers of the contract being tagged as money-containing. The goal field represents the amount of money the owner of the contract is trying to raise, rather than an amount of money owned by the contract. However, the field is still tagged as "money," since its value is regularly compared to the value of balance. As the backers field is a map, the analysis determines that its values represent money.

In a case where money-related fields are handled inconsistently (e.g., the value of goal is added to deadline), the analysis will report an error. This treatment of asset-related fields via the static analysis is similar to the idea of inferring *units of measure* as auxiliary information for types in a program [Kennedy 1997].

Type systems for managing digital assets

Static safety of asset management is an excellent application for type-based techniques. A particularly prominent idea is to use *linear types* [Girard 1987, Wadler 1993]—a form of type system that ensures that values that a program operates with are always "consumed" exactly once. Linear types have been previously employed to control resource usage in functional programs [Turner et al. 1995]. As such, linear types are also a good fit to describe the data type of assets. This way, the crucial property of "no double spending" of digital assets will be enforced by the type discipline, as a double spending would be similar to "double usage," which is precisely what linear types prevent from happening.

Linear types were first used in the Typecoin system [Crary and Sullivan 2015] to ensure the absence of double-spend Bitcoin scripts [Bitcon Wiki 2017]. A similar idea has been used in the Nomos language [Das et al. 2019], in which linear types were integrated with session types, and were used to define the consistent money transfer as a part of statically enforced communication protocols. Both Flint [Schrans 2018] and Obsidian [Coblenz et al. 2019] languages define a form of Asset type whose values obey the linearity property. Flint's notion of assets prevents the values of that type from being duplicated or destroyed accidentally. This choice leads to certain limitations on how assets are handled in Flint: for example, they cannot be returned from functions. Obsidian addresses these issues by instrumenting its types with *access permissions*, controlling the ways certain values are used. This way, it permits any non-primitive type to be an asset and treated as a first-class value.

Move [Blackshear et al. 2019] is a statically typed stack-based bytecode language with a syntactic layer providing an intermediate representation that is sufficiently high-level to write human-readable code, yet directly translates to Move bytecode. The key feature of Move's type system is the ability to define custom resource types with semantics inspired by linear logic [Girard 1987]: a resource can never be copied or implicitly discarded, only moved between program storage locations. Move's programming model assumes a global mapping of addresses representing the blockchain entities to assets (resources). Therefore, anyone can change their account by publishing new resources representing "currencies" of different kinds. Linearity of resources ensures that users cannot lose or duplicate their assets when transforming them from one kind to another. In contrast to Move, Scilla does not have a notion of a global mutable mapping from addresses to assets. This implies that contract authors commonly have to maintain their own local mappings of addresses to assets tailored to the purpose of the contract at hand. Handling those local mappings can often be simplified by using a generic escrow-like contract published alongside the specific contract the programmer wants to deploy [Trunov 2019].

3.6   EXECUTION COSTS AND GAS ACCOUNTING

The rationale behind the resource-aware smart contract semantics, instrumented with gas consumption, is three-fold. First, paying for gas at the moment of proposing the transaction does not allow the emitter to waste other parties' *computational power* by requiring them to perform a lot of worthless intensive work. Second, gas fees disincentivize users to consume too much *replicated storage*, which is a valuable resource in a blockchain-based consensus system. Finally, such a semantics puts a cap on the number of computations that a transaction can execute, and hence prevents denial-of-service attacks based on non-terminating executions (which could otherwise, for example, make all miners loop forever).

EVM provides a detailed specification of gas cost allocations for all its primitive commands [Wood 2014], with a lot of focus on costs for interacting with the storage. For instance, EVM's memory model defines three areas where it can store items: (a) the *storage* is where all contract state variables reside—every contract has its own storage, and it is persistent between external function calls (transactions) and quite expensive to use; (b) the *memory* is used to hold temporary values—it is erased between transactions and is cheaper to use than storage; (c) the *stack* is used to carry out operations and is free to use, but it can only hold a limited amount of values.

The gas-aware operational semantics of EVM has introduced novel challenges w.r.t. sound static reasoning about resource consumption, correctness, and security of replicated computations:

1. Ethereum safety recommendations discourage having the gas consumption of smart contracts depend on either the size of the data it stores (i.e., the contract state), the size of its functions' inputs, or the current state of the blockchain [E. Foundation 2018]. However, according to a recent study, almost 10% of the functions of Ethereum contracts feature such a dependency [Albert et al. 2019]. The inability to estimate those dependencies, along with the lack of analysis tools, leads to design mistakes that make a contract unsafe to run or prone to exploits. For instance, a contract whose state size exceeds a certain limit can be made to be forever stuck, not able to perform any operation within a reasonable gas bound. Those vulnerabilities have been recognized before, but only discovered by means of unsound, pattern-based analysis [Grech et al. 2018].

2. While the EVM specification enumerates the precise gas consumption for low-level operations, most of the smart contracts are written in high-level languages, such as Solidity [Ethereum Foundation 2019] or Vyper [Ethereum Foundation 2018b]. The translation of the high-level language constructs to the low-level ones makes static estimation of runtime

gas bounds challenging, and is implemented in an ad hoc way by state-of-the-art compilers, which are only able to give constant gas bounds, or return ∞ otherwise.

3.6.1   CONTROLLING GAS CONSUMPTION WITH PL TECHNIQUES

The programming language research to date has addressed the first of these challenges by applying a number of techniques to support more accurate gas consumption analysis. One of the best studied approaches for enforcing non-functional properties such as resource consumption is by employing *sub-structural type systems*, allowing programmers to declare the desired boundaries on resources consumed by the program and letting the type checker (as a part of the compilation pipeline) ensure statically that these boundaries are respected at runtime [Hoffmann and Shao 2014, Wang et al. 2017]. As writing explicit resource boundaries might impose large annotation overhead, thus slowing down the contract development, type-based techniques are usually combined with a static analysis that facilitates the inference of resource boundaries [Hoffmann et al. 2017]. Recent approaches have combined session types [Honda et al. 1998] with automated type-based resource inference [Das et al. 2019], making the resource consumption become part of an interaction protocol between different contracts and their users.

Gas analysis in account/balance model

The primary goal of estimating the execution cost (in terms of gas) for a transaction involving a smart contract is to predict the amount of digital currency one needs to pay to miners for processing the transaction. However, as a recent study shows, in an *account/balance* blockchain model such costs are typically parametric in the values of the transaction parameters as well as in the values of certain blockchain components, which might not be known at the moment when the transaction is broadcast [Sergey et al. 2019]. As a simple example, imagine a contract whose execution depends on the value of a block number in which the transaction is going to be adopted—or on the value of another contract's state component that will have changed between the moment the transaction is proposed and the moment it is processed.

Therefore, contrary to the common perception that the main virtue of a sound and complete gas analyzer for smart contracts is to predict the *exact* dynamic gas consumption [Marescotti et al. 2018, Wang 2019], we believe the main benefit of such an analysis is the possibility of detecting gas inefficiency patterns prior to contract deployment [Chen et al. 2017]. For instance, the gas analyzer shipped along with the Scilla programming language may return the $\top$ result when it fails to infer a polynomial boundary [Sergey et al. 2019]. Assuming the soundness of the analysis, though, even this $\top$ result of the analysis is still informative, as it indicates worse-than-linear gas consumption, which is usually a design flaw.

Gas analysis in UTXO model

In comparison with the account/balance model, gas analysis in the UTXO model is relatively straightforward, and is much more precise, as it is able to provide the *exact* execution cost. This is due to the fact that in UTXO each transaction, when proposed, should specify its "predecessor" transactions in the adopted history so far, thus fixing its input values at that moment. The obvious downside of such an execution model is a potential bottleneck: if several transactions that depend on the same output are proposed concurrently, at most one of them will be adopted. In contrast, the non-determinism of the account/balance model allows all concurrently proposed non-conflicting transactions to be committed within the same block.

3.6.2   GAS CONSUMPTION AND COMPILATION

Gas costs need to be stated in terms of a certain operational model and fixed for each execution unit. In this regard, EVM's design is the most uncontroversial, as it puts gas costs on the most primitive commands. However, this complicates the gas analysis for any of the high-level languages that are compiled to EVM. Not only are they imprecise, but modification and optimization in a compiler pipeline might break their analysis logic, which will have to be adapted accordingly.

What if the high-level contract language is what's being deployed on the blockchain? Such a design decision—adopted, for instance, by SCILLA—has made it possible to state the gas boundaries in terms of the operational semantics describing the evaluation of contracts using the high-level language [Sergey et al. 2019]. This choice can potentially face an unexpected threat: the high-level gas boundaries can be invalidated if an efficient optimizing compiler is implemented that significantly reduces the projected execution costs of certain commands.

That said, even though EVM is a low-level language, it is prone to a similar potential issue, as it is no longer directly interpreted, but rather is compiled *just-in-time* as a contract is being executed [E Foundation 2019]; hence the gas costs defined by its specification might be rendered inadequate. As another recent case study shows, this conjecture has been proven in practice, and the current ill-defined gas costs in Ethereum open it up to certain kinds of denial-of-service attacks [Pérez and Livshits 2019].

One of the most promising approaches to balance the discrepancy between the high-level and the low-level gas consumption is to employ the idea of typed assembly language (TAL; see Morrisett et al. [1998]) and type-preserving compilation. Wang's PhD thesis [Wang 2019] explores this direction and achieves very promising results showing that one can have high-level resource boundaries accurately translated to the boundaries for low-level code. The main disadvantages of the TAL-based resource analysis and compilation are (a) the requirement to have an expressive enough (and usually very hard-to-design) type system for the target language and (b) the need to preserve the typing information across compiler optimizations.

3.7   STANDING RESEARCH PROBLEMS

Before concluding, we provide a list of what we believe are important (and yet unsolved) problems in the programming language design for smart contracts.

Equivalence of account/balance and UTXO models

The choice of language abstractions and static reasoning mechanisms is determined to a large extent by the state model of the underlying consensus protocol: UTXO or account/balance. While the former provides better opportunities for parallel execution and encourages functional programming style, it makes it difficult to implement an arbitrary state and, specifically, encode custom currencies in addition to the native one. That said, to date it is unclear whether the choice of the underlying execution model impacts the expressivity of the application layer—in other words, the conjecture of whether the same set of contract-enabled applications can be encoded on top of both UTXO and the account/balance model.

Equivalence of bank-centric and account-centric models

MOVE's approach to managing assets is different from those of Solidity, SCILLA, Obsidian, and Flint in the sense that it allocates the resources (i.e., the amounts of custom currencies) right along with the accounts they belong to and piggybacks on the runtime's mechanism to ensure the absence of duplication. In contrast, other languages' approaches force each contract that introduces a custom notion of a token to provide the functionality that controls the distribution of the tokens between the users of the contract. Two models, therefore, appear to be complementary: in the account-centric model, the currency definition provides the rules only for splitting and joining the funds, while the virtual machine takes care of enforcing the linearity; however, in the contract-centric model, all the accounting logic is implemented by the contracts themselves, which play the roles of independent "banks." A theoretical question to be answered is whether these two models are equivalent in terms of expressivity, and also with regard to their desired security properties—and, if so, whether one can automatically translate digital contracts implemented in one of these paradigms into those of another one.

Adequate accounting of gas costs

We believe that the gas-aware nature of smart contract programming poses some of the most interesting unsolved challenges in PL design and implementation.

The first problem that needs to be solved is addressing the discrepancies between the high-level and low-level languages in terms of the projected gas consumption and the definitions of the gas costs. While we mention some ways to achieve that in Section 3.6.2, they are by no means a final solution. Specifically, they cannot account for the specialized hardware that might be employed by certain miners to decrease the intrinsic cost of transaction processing, thus giving them some unfair advantage. Therefore, the ideal gas costs need to be *amortized* with regard to multiple ways of executing the same contract commands.

The other challenge is to extend the gas allocation to other aspects of contract-related transaction processing. For instance, EVM only charges gas for dynamic contract execution, while SCILLA also mandates the miners to type-check the contracts upon deployment, as well as validate all the messages for the type information. This imposes additional intrinsic transaction processing costs, which do not fit the gas model of the smart contract language, yet need to be enumerated for the same reasons the gas price is put on contract executions. In the future, we foresee more validation checks will need to be added to the mining routine. This is why a principled (non–ad hoc) solution for converting *intrinsic* execution costs of mining to the *extrinsic* gas prices is very much desired.

3.8 CONCLUSION

Peter Landin's seminal paper "The Next 700 Programming Languages" [Landin 1966] stated the following seminal thesis: ". . . we must systematize [language] design so that a new language is a point chosen from a well-mapped space, rather than a laboriously devised construction." This thesis is indeed applicable to the specific family of programming languages used for implementing digital contracts.

We have surveyed the essential concepts of smart contract intrinsics: *atomicity*, *communication*, *asset management*, and *resource accounting*. We believe that any language for smart contracts coming in the future will have to provide suitable abstractions for expressing and manipulating these concepts, which serve as a set of basic building blocks for constructing reliable and trustworthy blockchain applications.

BIBLIOGRAPHY

æternity Blockchain. Sophia, 2019. https://github.com/aeternity/protocol/blob/master/contracts/sophia .md. 72

G. A. Agha. *ACTORS—A Model of Concurrent Computation in Distributed Systems*. MIT Press Series in Artificial Intelligence. MIT Press, 1990. 81

M. Al-Bassam, A. Sonnino, S. Bano, D. Hrycyszyn, and G. Danezis. Chainspace: A sharded smart contracts platform. In *25th Annual Network and Distributed System Security (NDSS) Symposium 2018, San Diego, California, USA, February 18–21, 2018*, pages 1–15. The Internet Society, 2018. 71

E. Albert, P. Gordillo, A. Rubio, and I. Sergey. Running on fumes—Preventing out-of-gas vulnerabilities in Ethereum smart contracts using static resource analysis. In *13th International Conference on Verification and Evaluation of Computer and Communication Systems (VECoS), Porto, Portugal*, volume 11847 of *Lecture Notes in Computer Science (LNCS)*, pages 63–78. Springer, 2019. 70, 84

J. Aldrich, J. Sunshine, D. Saini, and Z. Sparks. Typestate-oriented programming. In *24th Annual ACM SIGPLAN Conference on Object-Oriented Programming, Systems, Languages, and Applications (OOPSLA 2009), Orlando, Florida, USA*, pages 1015–1022. ACM, 2009. 80

G. Alfour. Introducing LIGO: A new smart contract language for Tezos, 2019. https://medium.com/tezos/introducing-ligo-a-new-smart-contract-language-for-tezos-233fa17f21c7. 72

J. Alois. Ethereum parity hack may impact ETH 500,000 or $146 million, 2017. https://www.crowdfundinsider.com/2017/11/124200-ethereum-parity-hack-may-impact-eth-500000-146-million/. 70

L. Alt and C. Reitwießner. SMT-based verification of Solidity smart contracts. In *8th International Symposium on Leveraging Applications of Formal Methods, Verification and Validation (ISoLA 2018), Limassol, Cyprus*, volume 11247 of *LNCS*, pages 376–388. Springer, 2018. 72

S. Amani, M. Bégel, M. Bortin, and M. Staples. Towards verifying Ethereum smart contract bytecode in Isabelle/HOL. In *7th ACM SIGPLAN International Conference on Certified Programs and Proofs (CPP 2018), Los Angeles, CA, USA*, pages 66–77. ACM, 2018. 72

M. D. Angelo and G. Salzer. A survey of tools for analyzing Ethereum smart contracts. In *IEEE International Conference on Decentralized Applications and Infrastructures, DAPPCON*, pages 69–78. IEEE, 2019. 72

J. Armstrong. A history of Erlang. In *Proceedings of the 3rd ACM SIGPLAN History of Programming Languages Conference (HOPL-III)*, pages 1–26. ACM, 2007. 81

N. Atzei, M. Bartoletti, and T. Cimoli. A survey of attacks on Ethereum smart contracts (SoK). In *6th International Conference on Principles of Security and Trust (POST 2017), Uppsala, Sweden*, volume 10204 of *LNCS*, pages 164–186. Springer, 2017. 70

S. Bano, A. Sonnino, M. Al-Bassam, S. Azouvi, P. McCorry, S. Meiklejohn, and G. Danezis. SoK: Consensus in the age of blockchains. In *Proceedings of the 1st ACM Conference on Advances in Financial Technologies (AFT 2019), Zurich, Switzerland*, pages 183–198, 2019. 69

K. Bansal, E. Koskinen, and O. Tripp. Automatic generation of precise and useful commutativity conditions. In *24th International Conference on Tools and Algorithms for the Construction and Analysis of Systems (TACAS 2018), Thessaloniki, Greece*, volume 10805 of *LNCS*, pages 115–132. Springer, 2018. 72

K. Bhargavan, A. Delignat-Lavaud, C. Fournet, A. Gollamudi, G. Gonthier, N. Kobeissi, N. Kulatova, A. Rastogi, T. Sibut-Pinote, N. Swamy, and S. Zanella-Béguelin. Formal verification of smart contracts: Short paper. In *11th Workshop on Programming Languages and Analysis for Security (PLAS 2016), Vienna, Austria*, pages 91–96. ACM, 2016. 72

Bitcoin Wiki. Bitcoin Script, 2017. https://en.bitcoin.it/wiki/Script. (Accessed 5 Apr. 2019.) 83

S. Blackshear, E. Cheng, D. L. Dill, V. Gao, B. Maurer, T. Nowacki, A. Pott, S. Qadeer, Rain, D. Russi, S. Sezer, T. Zakian, and R. Zhou. MOVE: A language with programmable resources, 2019. https://developers.libra.org/docs/assets/papers/libra-move-a-language-with-programmable-resources.pdf. 83

V. Buterin. A next generation smart contract & decentralized application platform, 2013. https://www.ethereum.org/pdfs/EthereumWhitePaper.pdf/. 69, 70

M. M. T. Chakravarty, J. Chapman, K. MacKenzie, O. Melkonian, M. P. Jones, and P. Wadler. The extended UTXO model. *Financial Cryptography and Data Security (FC 2020)*, volume 12063 of *LNCS*, pages 525–539. Springer, 2020. 77

J. Chang, B. Gao, H. Xiao, J. Sun, and Z. Yang. sCompile: Critical path identification and analysis for smart contracts. *CoRR*, abs/1808.00624, 2018. 72

T. Chen, X. Li, X. Luo, and X. Zhang. Under-optimized smart contracts devour your money. In *IEEE 24th International Conference on Software Analysis, Evolution and Reengineering, SANER*, pages 442–446. IEEE Computer Society, 2017. 70, 85

M. Coblenz. Obsidian: A safer blockchain programming language. In *39th International Conference on Software Engineering (ICSE 2017), Buenos Aires, Argentina*, pages 97–99. IEEE Press, 2017. 72

M. J. Coblenz, R. Oei, T. Etzel, P. Koronkevich, M. Baker, Y. Bloem, B. A. Myers, J. Sunshine, and J. Aldrich. Obsidian: Typestate and assets for safer blockchain programming. *CoRR*, abs/1909.03523, 2019. 80, 83

Coq Development Team. *The Coq Proof Assistant Reference Manual—Version 8.10*, 2019. http://coq.inria.fr. 76

P. Cousot and R. Cousot. Abstract interpretation: A unified lattice model for static analysis of programs by construction or approximation of fixpoints. In *4th ACM Symposium on Principles of Programming Languages (POPL 1977), Los Angeles, CA, USA*, pages 238–252. ACM, 1977. 82

K. Crary and M. J. Sullivan. Peer-to-peer affine commitment using Bitcoin. In *36th ACM SIGPLAN Conference on Programming Language Design and Implementation (PLDI 2015), Portland, OR, USA*, pages 479–488. ACM, 2015. 83

A. Das, S. Balzer, J. Hoffmann, and F. Pfenning. Resource-aware session types for digital contracts. *CoRR*, abs/1902.06056, 2019. 82, 83, 85

A. Das, J. Hoffmann, and F. Pfenning. Work analysis with resource-aware session types. In *33rd Annual ACM/IEEE Symposium on Logic in Computer Science (LICS 2018), Oxford, UK*, pages 305–314. ACM, 2018. 82

M. del Castillo. The DAO attacked: Code issue leads to $60 million Ether theft, 2016. https://www.coindesk.com/dao-attacked-code-issue-leads-60-million-ether-theft/. (Accessed 2 Dec. 2017.) 70, 76, 78

C. Dong, Y. Wang, A. Aldweesh, P. McCorry, and A. van Moorsel. Betrayal, distrust, and rationality: Smart counter-collusion contracts for verifiable cloud computing. In *24th ACM SIGSAC Conference on Computer and Communications Security (CCS 2017), Dallas, TX, USA*, pages 211–227. ACM, 2017. 71

Ethereum Foundation. Decentralized Autonomous Organization, 2018a. https://www.ethereum.org/dao. 70

Ethereum Foundation. Vyper, 2018b. https://vyper.readthedocs.io. 84

Ethereum Foundation. Solidity documentation, 2019. http://solidity.readthedocs.io. 74, 84

E. Foundation. Safety—Ethereum Wiki, 2018. https://github.com/ethereum/wiki/wiki/Safety. 84

E. Foundation. The Ethereum EVM JIT, 2019. https://github.com/ethereum/evmjit. 86

J. Girard. Linear logic. *Theor. Comput. Sci.* 50:1–102, 1987. 83

N. Grech, M. Kong, A. Jurisevic, L. Brent, B. Scholz, and Y. Smaragdakis. MadMax: Surviving out-of-gas conditions in Ethereum smart contracts. *Proceedings of the ACM on Programming Languages (PACMPL), Volume 2 (OOPSLA)*:116:1–116:27, 2018. 70, 72, 84

I. Grishchenko, M. Maffei, and C. Schneidewind. A semantic framework for the security analysis of Ethereum smart contracts. In *7th International Conference on Principles of Security and Trust (POST 2018)*, volume 10804 of *LNCS*, pages 243–269. Springer, 2018. 72

S. Grossman, I. Abraham, G. Golan-Gueta, Y. Michalevsky, N. Rinetzky, M. Sagiv, and Y. Zohar. Online detection of effectively callback free objects with applications to smart contracts. *Proceedings of the ACM on Programming Languages (PACMPL)*, volume 2 (POPL), pages 1–28, 2018. https://doi.org/10.1145/3158136 72, 76

E. Gün Sirer. Reentrancy woes in smart contracts, 2016. https://hackingdistributed.com/2016/07/13/reentrancy-woes/. (Accessed Oct. 2020.) 76, 78

P. Haller and F. Sommers. *Actors in Scala—Concurrent Programming for the Multi-core Era*. Artima, 2011. 81

M. Herlihy and N. Shavit. *The Art of Multiprocessor Programming*. Morgan Kaufmann, 2008. 78

Y. Hirai. Bamboo, 2018. https://github.com/pirapira/bamboo. 72

J. Hoffmann, A. Das, and S. Weng. Towards automatic resource bound analysis for OCaml. In *POPL*, pages 359–373. ACM, 2017. 85

J. Hoffmann and Z. Shao. Type-based amortized resource analysis with integers and arrays. In *12th International Symposium on Functional and Logic Programming (FLOPS 2014), Kanazawa, Japan*, volume 8475 of *LNCS*, pages 152–168. Springer, 2014. 85

K. Honda, V. T. Vasconcelos, and M. Kubo. Language primitives and type discipline for structured communication-based programming. In *7th European Symposium on Programming (ESOP 1998), held as part of the European Joint Conferences on the Theory and Practice of Software (ETAPS'98), Lisbon, Portugal*, volume 1381 of *LNCS*, pages 122–138. Springer, 1998. 85

IOHK Foundation. Marlowe: A contract language for the financial world, 2019a. https://testnet.iohkdev.io/marlowe/. 72

IOHK Foundation. Plutus: A functional contract platform, 2019b. https://testnet.iohkdev.io/plutus/. 80

J. Johannsen and A. Kumar. Introducing the ZIL Cashflow Smart Contract Analyser, 2019. Blog post available at https://blog.zilliqa.com/introducing-the-zil-cashflow-smart-contract-analyser-ded8b4d84362. 82

S. Kalra, S. Goel, M. Dhawan, and S. Sharma. Zeus: Analyzing safety of smart contracts. In *25th Annual Network and Distributed System Security Symposium (NDSS 2018), San Diego, California, USA*, 2018. 70, 72, 76

A. Kennedy. Relational parametricity and units of measure. In *POPL*, pages 442–455. ACM Press, 1997. 83

A. Kiayias, A. Russell, B. David, and R. Oliynykov. Ouroboros: A provably secure proof-of-stake blockchain protocol. In *CRYPTO, Part I*, volume 10401 of *LNCS*, pages 357–388. Springer, 2017. 77

E. Kokoris-Kogias, P. Jovanovic, L. Gasser, N. Gailly, E. Syta, and B. Ford. OmniLedger: A secure, scale-out, decentralized ledger via sharding. In *2018 IEEE Symposium on Security and Privacy (SP)*, pages 583–598. IEEE Computer Society, 2018. 71

A. Kolluri, I. Nikolic, I. Sergey, A. Hobor, and P. Saxena. Exploiting the laws of order in smart contracts. In *28th ACM SIGSOFT International Symposium on Software Testing and Analysis (ISSTA 2019), Beijing, China*, pages 363–373. ACM, 2019. 70, 72

J. Krupp and C. Rossow. teEther: Gnawing at Ethereum to automatically exploit smart contracts. In *USENIX Security Symposium*, pages 1317–1333. USENIX Association, 2018. 70, 72

P. J. Landin. The next 700 programming languages. *Commun. ACM* 9(3):157–166, 1966. 88

L. Luu, D. Chu, H. Olickel, P. Saxena, and A. Hobor. Making smart contracts smarter. In *CCS*, pages 254–269. ACM, 2016a. 69, 72

L. Luu, V. Narayanan, C. Zheng, K. Baweja, S. Gilbert, and P. Saxena. A secure sharding protocol for open blockchains. In *CCS*, pages 17–30. ACM, 2016b. 71

N. A. Lynch and M. R. Tuttle. An introduction to input/output automata. *CWI Quarterly* 2:219–246, 1989. 81

M. Marescotti, M. Blicha, A. E. J. Hyvärinen, S. Asadi, and N. Sharygina. Computing exact worst-case gas consumption for smart contracts. In *ISoLA*, volume 11247 of *LNCS*, pages 450–465. Springer, 2018. 70, 72, 85

P. McCorry, A. Hicks, and S. Meiklejohn. Smart contracts for bribing miners. In *Financial Cryptography and Data Security—FC 2018 International Workshops*, volume 10958 of *LNCS*, pages 3–18. Springer, 2019. 71

P. McCorry, S. F. Shahandashti, and F. Hao. A smart contract for boardroom voting with maximum voter privacy. In *FC*, volume 10322 of *LNCS*, pages 357–375. Springer, 2017. 71

R. Milner. *A Calculus of Communicating Systems*, volume 92 of *Lecture Notes in Computer Science*. Springer, 1980. 81

J. G. Morrisett, D. Walker, K. Crary, and N. Glew. From System F to typed assembly language. In *POPL*, pages 85–97. ACM, 1998. 86

S. Nakamoto. Bitcoin: A peer-to-peer electronic cash system, 2008. Available at http://bitcoin.org/bitcoin.pdf. 69, 77

I. Nikolić, A. Kolluri, I. Sergey, P. Saxena, and A. Hobor. Finding the greedy, prodigal, and suicidal contracts at scale. In *34th Annual Computer Security Applications Conference (ACSAC 2018), San Juan, PR, USA*, pages 653–663. ACM, 2018. 70, 72, 82

OCaml PRO. Liquidity, 2019. https://www.liquidity-lang.org/. 77

R. O'Connor. Simplicity: A new language for blockchains, 2017. https://blockstream.com/simplicity.pdf. 80

D. Pérez and B. Livshits. Broken metre: Attacking resource metering in EVM. *CoRR*, abs/1909.07220, 2019. 70, 86

A. Permenev, D. Dimitrov, P. Tsankov, D. Drachsler-Cohen, and M. Vechev. VerX: Safety verification of smart contracts. In *IEEE Symposium on Security and Privacy SP*, 2020. 76, 77

B. C. Pierce. *Types and Programming Languages*. MIT Press, 2002. 78

G. Pîrlea and I. Sergey. Mechanising blockchain consensus. In *CPP*, pages 78–90. ACM, 2018. 69

RChain Cooperative. Rholang, 2019. https://rholang.rchain.coop. 72, 82

C. Reitwiessner. Babbage—A mechanical smart contract language, 2017. Online blog post. 72

M. Rodler, W. Li, G. O. Karame, and L. Davi. Sereum: Protecting existing smart contracts against re-entrancy attacks. In *NDSS*, 2019. 70

F. Schrans. *Writing Safe Smart Contracts in Flint*. Master's thesis, Imperial College London, Department of Computing, 2018. 80, 83

F. Schrans, S. Eisenbach, and S. Drossopoulou. Writing safe smart contracts in Flint. In *<Programming> (Companion)*, pages 218–219. ACM, 2018. 80

Securify. https://securify.chainsecurity.com/. (Accessed 2 Jan. 2019.) 76

P. L. Seijas, S. J. Thompson, and D. McAdams. Scripting smart contracts for distributed ledger technology. *IACR Cryptology ePrint Archive*, 2016. 72

I. Sergey and A. Hobor. A concurrent perspective on smart contracts. In *1st Workshop on Trusted Smart Contracts (WTSC 2017), Malta*, volume 10323 of *LNCS*, pages 478–493. Springer, 2017. 78, 79, 80

I. Sergey, A. Kumar, and A. Hobor. SCILLA: a Smart Contract Intermediate-Level LAnguage. *CoRR*, abs/1801.00687, 2018a. 76, 81

I. Sergey, A. Kumar, and A. Hobor. Temporal properties of smart contracts. In *ISoLA*, volume 11247 of *LNCS*, pages 323–338. Springer, 2018b. 77

I. Sergey, V. Nagaraj, J. Johannsen, A. Kumar, A. Trunov, and K. C. G. Hao. Safer smart contract programming with SCILLA. *PACMPL*, 3(OOPSLA):185:1–185:30, 2019. 77, 79, 80, 85, 86

F. Sun. UTXO vs Account/Balance Model, 2018. Online blog post, available at https://medium.com/@sunflora98/utxo-vs-account-balance-model-5e6470f4e0cf. 77

N. Szabo. Smart contracts, 1994. Online manuscript. 69

Tezos Foundation. Michelson: The Language of Smart Contracts in Tezos, 2018. http://tezos.gitlab.io/mainnet/whitedoc/michelson.html. 77

S. Tikhomirov, E. Voskresenskaya, I. Ivanitskiy, R. Takhaviev, E. Marchenko, and Y. Alexandrov. SmartCheck: Static analysis of Ethereum smart contracts. In *1st IEEE/ACM International Workshop on Emerging Trends in Software Engineering for Blockchain (WETSEB@ICSE 2018), Gothenburg, Sweden*, pages 9–16. ACM, 2018. 72

A. Trunov. A SCILLA vs MOVE case study, 2019. Blog post available at https://medium.com/@anton_trunov/a-scilla-vs-move-case-study-afa9b8df5146. 84

P. Tsankov, A. M. Dan, D. Drachsler-Cohen, A. Gervais, F. Bünzli, and M. T. Vechev. Securify: Practical security analysis of smart contracts. In *CCS*, pages 67–82. ACM, 2018. 72, 76

D. N. Turner, P. Wadler, and C. Mossin. Once upon a type. In *7th International Conference on Functional Programming Languages and Computer Architecture (FPCA 1995), La Jolla, CA, USA*, pages 1–11. ACM, 1995. 83

P. Wadler. A taste of linear logic. In *Mathematical Foundations of Computer Science 1993, 18th International Symposium, MFCS'93*, volume 711 of *LNCS*, pages 185–210. Springer, 1993. 83

P. Wang. *Type System for Resource Bounds with Type-Preserving Compilation*. PhD thesis, Massachusetts Institute of Technology, 2019. 85, 86

P. Wang, D. Wang, and A. Chlipala. TiML: A functional language for practical complexity analysis with invariants. *PACMPL*, 1(OOPSLA):79:1–79:26, 2017. 85

G. Wood. Ethereum: A secure decentralized generalized transaction ledger, 2014. Ethereum Project Yellow Paper, 2014. https://github.com/ethereum/yellowpaper. 69, 70, 84

Zilliqa Team. The ZILLIQA technical whitepaper, 2017. Version 0.1. 81

AUTHOR'S BIOGRAPHY

Ilya Sergey is an Associate Professor at Yale-NUS College and School of Computing of National University of Singapore. He also serves as a lead language designer at Zilliqa, a Singapore FinTech start-up. Before joining NUS in 2018, Ilya was a faculty member at University College London, a postdoctoral researcher at IMDEA Software Institute, and a software developer at JetBrains. He completed his PhD at KU Leuven in 2012. Ilya does research in programming language design and implementation, software verification, program synthesis and repair. He is the recipient of the 2019 Dahl-Nygaard Junior Prize and the Google Faculty Research Award 2017. He designed and co-developed SCILLA, a functional programming language for safe smart contracts, used by ZILLIQA blockchain. Ilya was an organizer of the ICFP Programming Contest in 2019, and will serve as a Programme Committee Chair for ESOP'22 and APLAS'22.

CHAPTER 4

Formalization of
Blockchain Properties

Emmanuelle Anceaume, *CNRS, Univ. Rennes, IRISA, France*
Antonio Fernández Anta, *IMDEA Networks Institute, Madrid, Spain*
Chryssis Georgiou, *University of Cyprus, Nicosia, Cyprus*
Nicolas Nicolaou, *Algolysis Ltd, Limassol, Cyprus*
Maria Potop-Butucaru, *LIP6, Sorbonne University, Paris, France*

4.1   INTRODUCTION

There is no doubt that cryptocurrencies and (public and private) distributed ledgers (blockchains) have the potential to impact our society deeply. However, most experts often do not clearly differentiate between the coin, the ledger that supports it, and the service both provide. Instead, they get very technical, talking about the cryptography involved, the mining used to maintain the ledger, or the smart contract technology used. Moreover, when asked for details, it is often the case that there is no formal specification of the protocols, algorithms, and service provided, with a few exceptions (see related work below). In many cases "the code is the spec."

From the theoretical point of view there are many fundamental questions with the current distributed ledger (and cryptocurrency) systems that are very often not properly answered: What is the service that must be provided by a distributed ledger? What properties must a distributed ledger satisfy? What are the assumptions made by the protocols and algorithms on the underlying system? What are the consistency guarantees provided, given that many entities access the ledger concurrently? Does a distributed ledger require a linked cryptocurrency?

Recently, it was pointed out that despite the hype about blockchains and distributed ledgers, no formal abstraction of these objects has been proposed [Herlihy 2017]. In particular, it was stated that there is a need for the formalization of the distributed systems that are at the heart of most cryptocurrency implementations, and a need to leverage the decades of experience in the distributed computing community in formal specification when designing and proving various properties of such systems.

In this chapter, we present recent attempts to formalize the properties of distributed ledgers/ blockchains, from a distributed computing point of view. This is achieved by using the notion of *abstract data type*, which is a common abstraction to specify shared objects. The abstract data type of a shared object describes first the semantics of the object by providing its sequential specification and, second, its consistency criterion, which describes its behavior in the presence of concurrent accesses. This notion of abstract data type is presented in Section 4.2. Then in Section 4.3 we present the concept of a *distributed ledger object* [Fernández Anta et al. 2018], which abstracts the main operations and properties of distributed ledgers, focusing mainly on permissioned systems; the presentation of this section is based on the one in Fernández Anta et al. [2018]. In Section 4.4, we present the concept of the *blockchain abstract data type* [Anceaume et al. 2019b], which provides a lower-level abstraction of distributed ledgers, suitable for both permissioned and permissionless systems; the presentation of this section is based on the one in Anceaume et al. [2019b]. The chapter concludes with a discussion in Section 4.5.

Related work

The first effort in specifying the properties of permissionless blockchain systems is due to Garay et al. [2015]. They characterize the Bitcoin blockchain via its quality and its common prefix properties; specifically, they define an invariant that this protocol has to satisfy in order to verify with high probability an eventually consistent prefix. This line of work has been continued by Pass and Shi [2017].

In order to model the behavior of distributed ledgers at runtime, Girault et al. [2018] present an implementation of the *monotonic prefix consistency* (MPC) criterion and show that no criterion stronger than MPC can be implemented in a partition-prone message-passing system. On the other hand, the proposed formalization does not suggest weaker consistency semantics more suitable for proof-of-work blockchains such as Bitcoin.

The work in Anceaume et al. [2017] makes a connection between the distributed ledgers and the distributed shared objects (registers) theory. The authors introduce the notion of a *distributed ledger register* (DLR) where the value of the register has a tree topology instead of a single value as in the classical theory of distributed registers. The vertices of the tree are blocks of transactions linked cryptographically. The DLR properties were crafted to fit the behavior of permissionless blockchains such as Bitcoin and Ethereum. The work presented in Section 4.3 defines the ledger object as an ordered sequence of records, abstracting away from registers. The work presented in Section 4.4 also moves away from registers and refines the specifications of Anceaume et al. [2017] in order to cover both weaker semantics (i.e., safe shared objects) and stronger semantics (i.e., atomic shared objects).

4.2   PRELIMINARIES

Abstract data types can be used to specify shared objects using two complementary facets [Perrin 2017]: a sequential specification that describes the semantics of the object, and a consistency criterion over concurrent histories—that is, the set of admissible executions in a concurrent environment.

Intuitively, an *abstract data type* T specifies (i) the set of *values* (or states) that any object O of type T can take, and (ii) the set of *operations* that a process (client) can use to modify or access the value of O. An object O of type T is a *concurrent object* if it can be accessed by multiple processes concurrently [Raynal 2013]. Each operation on an object O consists of an *invocation* event and a *response* event, which must occur in this order. A *history* of operations on O, denoted by H_O, is a sequence of invocation and response events, starting with an invocation event. (The sequence order of a history reflects the real time ordering of the events.) An operation π is *complete* in a history H_O if H_O contains both the invocation and the matching response of π, in this order. A history H_O is *complete* if it contains only complete operations; otherwise it is *partial* [Raynal 2013]. An operation π_1 *precedes* an operation π_2 (or π_2 *succeeds* π_1), denoted by $\pi_1 \rightarrow \pi_2$, in H_O, if the response event of π_1 appears before the invocation event of π_2 in H_O. Two operations are *concurrent* if neither precedes the other.

A complete history H_O is *sequential* if it contains no concurrent operations—that is, it is an alternating sequence of matching invocation and response events, starting with an invocation and ending with a response event. A partial history is sequential if removing its last event (that must be an invocation) makes it a complete sequential history. A *sequential specification* of an object O describes the behavior of O when accessed sequentially. In particular, the sequential specification of O is the set of all possible sequential histories solely involving object O [Raynal 2013].

In the remainder of this section we provide formal definitions for the abstract data types, histories, and sequential specifications.

4.2.1   ABSTRACT DATA TYPE (ADT)

The model used to specify an abstract data type (ADT) is a form of transducer, as Mealy's machines—accepting an infinite but countable number of states. The values that can be taken by the data type are encoded in the abstract state, taken from a set Z. It is possible to access the object using the symbols of an input alphabet A. An operation can: (i) change the abstract state of an object using a transition function τ, and (ii) return values taken in an output alphabet B, which depend on the state in which they are called and an output function δ. For example, the pop operation in a stack removes the element at the top of the stack (changing its state) and returns that element (its output).

The formal definition of abstract data types is as follows.

Definition 4.1 (Abstract data type T)   An abstract data type is a 6-tuple $T = \langle A, B, Z, \xi_0, \tau, \delta \rangle$ where:

- A and B are countable sets called input alphabet and output alphabet;
- Z is a countable set of abstract states and ξ_0 is the initial abstract state;
- $\tau : Z \times A \to Z$ is the transition function;
- $\delta : Z \times A \to B$ is the output function.

Let α/β denote the input/output pair $(\alpha, \beta) \in A \times B$. We can define operations on an abstract data type as follows.

Definition 4.2 (Operation)   Let $T = \langle A, B, Z, \xi_0, \tau, \delta \rangle$ be an abstract data type. An *operation* of T is an element of $\Sigma = A \cup (A \times B)$.

We extend the transition and the output functions τ and δ on operations by applying τ and δ on the input symbol of the operations:

$$
\tau_T : \begin{cases} Z \times \Sigma \to Z \\ (\xi, \alpha) \mapsto \tau(\xi, \alpha) & \text{if } \alpha \in A \\ (\xi, \alpha/\beta) \mapsto \tau(\xi, \alpha) & \text{if } \alpha/\beta \in A \times B \end{cases}
$$

$$
\delta_T : \begin{cases} Z \times \Sigma \to Z \\ (\xi, \alpha) \mapsto \bot & \text{if } \alpha \in A \\ (\xi, \alpha/\beta) \mapsto \delta(\xi, \alpha) & \text{if } \alpha/\beta \in A \times B \end{cases}
$$

4.2.2   SEQUENTIAL SPECIFICATION OF AN ADT

An abstract data type defines the sequential specification of an object. That is, if we consider a path that traverses its system of transitions, then the word formed by the subsequent labels on the path is part of the sequential specification of the abstract data type; that is, it is a sequential history. The language recognized by an ADT is the set of all possible words. This language defines the sequential specification of the ADT. More formally, let $\delta_T^{-1}(\sigma_i)$ be the set of states of T in which a given operation can be realized:

$$
\delta_T^{-1}(\sigma_i) : \begin{cases} \Sigma \to \mathcal{P}(Z) \\ \alpha \mapsto Z & \text{if } \alpha \in A \\ \alpha/\beta \mapsto \{\xi \in Z : \delta(\xi, \alpha) = \beta\} & \text{if } \alpha/\beta \in A \times B \end{cases}
$$

Then we can define the sequential specification of T as follows.

Definition 4.3 (Sequential specification $L(T)$)   A finite (resp. infinite) sequence $\sigma = (\sigma_i)_{i \in D} \in \Sigma^{\infty}$, where $D \in \{0, \ldots, |\sigma| - 1\}$ (resp. $D \in \mathbb{N}$), is a *sequential history* of an abstract data type T if there exists a sequence of the same length $(\xi_{i+1})_{i \in D} \in Z^{\infty}$ of states of T (where ξ_0 is the initial state) such that, for any $i \in D$,

- the output alphabet of σ_i is compatible with ξ_i: $\xi_i \in \delta_T^{-1}(\sigma_i)$;

- the execution of the operation σ_i is such that the state changed from ξ_i to ξ_{i+1}: $\tau_T(\xi_i, \sigma_i) = \xi_{i+1}$.

The *sequential specification* of T is the set of all its possible sequential histories $L(T)$.

4.2.3 CONCURRENT HISTORIES OF AN ADT

Concurrent histories are defined considering asymmetric event structures—that is, partial order relations among events executed by different processes [Perrin 2017].

Definition 4.4 (Concurrent history H)   A concurrent history is an n-tuple $H = \langle \Sigma, E, \Lambda, \mapsto, \prec, \nearrow \rangle$ where:

- $\Sigma = A \cup (A \times B)$ is a countable set of operations of the abstract data type.

- E is a countable set of events that contains all the ADT operations invocations and all ADT operation response events.

- $\Lambda : E \to \Sigma$ is a function that associates events to the operations in Σ.

- $\mapsto$ is the process order relation over the events in E. Two events in E are ordered by $\mapsto$ if and only if they are invoked by the same process.

- $\prec$ is the operation order, irreflexive order over the events of E. For each couple $(e, e') \in E^2$, if e is an operation invocation and e' is the response for the same operation, then $e \prec e'$; if e' is the invocation of an operation that occurred at time t' and e is the response of another operation that occurred at time t with $t < t'$, then $e \prec e'$.

- $\nearrow$ is the program order, irreflexive order over E. For each couple $(e, e') \in E^2$ with $e \neq e'$, if $e \mapsto e'$ or $e \prec e'$, then $e \nearrow e'$.

4.2.4 CONSISTENCY CRITERION

A consistency criterion characterizes which concurrent histories are admissible for a given abstract data type. It can be viewed as a function that associates a concurrent specification to abstract data types. Specifically,

Definition 4.5 (Consistency criterion C)   A consistency criterion is a function

$$C : \mathcal{T} \to \mathcal{P}(\mathcal{H})$$

where $\mathcal{T}$ is the set of abstract data types, $\mathcal{H}$ is a set of histories, and $\mathcal{P}(\mathcal{H})$ is the set of all finite subsets of $\mathcal{H}$.

Let $\mathcal{C}$ be the set of all the consistency criteria. An algorithm A_T implementing the ADT $T \in \mathcal{T}$ is C-consistent with respect to criterion $C \in \mathcal{C}$ if all the operations terminate and all the executions are C-consistent—that is, they belong to the set of histories $C(T)$.

4.3   DISTRIBUTED LEDGER OBJECTS

In this section we present the *distributed ledger object* (DLO) formalism as first introduced in Fernández Anta et al. [2018]. In particular, Section 4.3.1 presents the notion of a *ledger object* as a concurrent object that maintains a totally ordered sequence of records, and supports two operations: get(), which returns the sequence, and append(r), which appends record r in the sequence. The sequential specification of the ledger object enforces only one record sequence to exist at any given time; that is, it prevents forks. This makes the formalism more suitable for modeling permissioned blockchains.

Section 4.3.2 extends the definition of the ledger object to the distributed ledger object, a concurrent ledger object that is implemented in a distributed manner, involving clients and servers. This gives rise to the need for specifying *consistency* guarantees that can be provided at the interface between clients and the servers implementing the ledger. Here, four different levels of consistency are formally specified: eventual, sequential, causal, and atomic (linearizable) consistency.

Section 4.3.3 provides an implementation of a linearizable distributed ledger in an asynchronous distributed system with crash failures. The implementation utilizes a crash-tolerant Atomic Broadcast service. Implementations of DLOs with the other weaker consistency guarantees can be found in Fernández Anta et al. [2018], and of a Byzantine-tolerant linearizable DLO in Cholvi et al. [2020].

Finally, Section 4.3.4 presents the notion of a *validated ledger*, an extension of the ledger object in which specific semantics are imposed on the contents of the records stored in the ledger. In other words, the ledger is equipped with the capability of performing application-specific validation checks on a record before being appended in the sequence.

4.3.1   THE LEDGER OBJECT
We begin with the fundamental definition of a concurrent ledger object.

Definition 4.6 (Ledger type)   A *ledger* $\mathcal{L}$ is a concurrent object that stores a totally ordered sequence $\mathcal{L}.S$ of *records* and supports two operations (available to any process p): (i) $\mathcal{L}.\text{get}_p()$, which returns the sequence $\mathcal{L}.S$, and (ii) $\mathcal{L}.\text{append}_p(r)$ which appends record r to $\mathcal{L}.S$.

A *record* is a triple $r = \langle \tau, p, v \rangle$, where τ is a *unique* record identifier from a set $\mathcal{T}$, $p \in \mathcal{P}$ is the identifier of the process that created record r, and v is the data of the record drawn from an alphabet A. We will use $r.p$ to denote the id of the process that created record r; we similarly define $r.\tau$ and $r.v$. A process p invokes an $\mathcal{L}.\text{get}_p()$ operation to obtain the sequence $\mathcal{L}.S$ of records stored in the ledger object $\mathcal{L}$, and p invokes an $\mathcal{L}.\text{append}_p(r)$ operation to extend $\mathcal{L}.S$ with a new record r. Initially, the sequence $\mathcal{L}.S$ is empty.

Definition 4.7 (Sequential specification with strong prefix)   The *sequential specification* of a ledger $\mathcal{L}$ over the sequential history $H_{\mathcal{L}}$ is defined as follows. The value of the sequence $\mathcal{L}.S$ of the ledger is initially the empty sequence. If at the invocation event of an operation π in $H_{\mathcal{L}}$ the value of the sequence in ledger $\mathcal{L}$ is $\mathcal{L}.S = V$, then:

1. if π is an $\mathcal{L}.\text{get}_p()$ operation, then the response event of π returns V, and
2. if π is an $\mathcal{L}.\text{append}_p(r)$ operation, then at the response event of π, the value of the sequence in ledger $\mathcal{L}$ is $\mathcal{L}.S = V \parallel r$ (where $\parallel$ is the concatenation operator). The response event of π returns ACK.

Implementation of ledgers

Processes execute operations and instructions sequentially (i.e., we make the usual well-formedness assumption where a process invokes one operation at a time). A process p interacts with a ledger $\mathcal{L}$ by invoking an operation ($\mathcal{L}.\text{get}_p()$ or $\mathcal{L}.\text{append}_p(r)$), which causes a request to be sent to the ledger $\mathcal{L}$, and a response to be sent from $\mathcal{L}$ to p. The response marks the end of an operation and also carries the result of that operation (the exchange of request and responses between the process and the ledger is made explicit, to reveal the fact that the ledger is concurrent, i.e., accessed by several processes). The result of a get operation is a sequence of records, while the result of an append operation is a confirmation (ACK). This interaction from the point of view of the process p is depicted in Algorithm 4.1. A possible centralized implementation of the ledger that processes requests sequentially is presented in Algorithm 4.2 (each block **receive** is assumed to be executed in mutual exclusion). Figure 4.1 (left) abstracts the interaction between the processes and the ledger.

4.3.2   FROM LEDGER OBJECTS TO DISTRIBUTED LEDGER OBJECTS

In this section we extend the definition of the ledger object to the distributed ledger object, and present some of the levels of *consistency* guarantees that can be provided at the interface between

---

Algorithm 4.1: External interface (executed by a process p) of a ledger object $\mathcal{L}$

1 **function** $\mathcal{L}$.get( )
2   **send** request (GET) **to** ledger $\mathcal{L}$
3   **wait** response (GETRES, V) **from** $\mathcal{L}$
4   **return** V

5 **function** $\mathcal{L}$.append(r)
6   **send** request (APPEND, r) **to** ledger $\mathcal{L}$
7   **wait** response (APPENDRES, res) **from** $\mathcal{L}$
8 **return** res

Algorithm 4.2: Ledger $\mathcal{L}$ (centralized)

1 **Init:** $S \leftarrow \emptyset$

2 **receive** (GET) **from** process p
3   **send** response (GETRES, S) **to** p
4 **end receive**

5 **receive** (APPEND, r) **from** process p
6      $S \leftarrow S \parallel r$
7   **send** response (APPENDRES, ACK) **to** p
8 **end receive**

---

processes (clients) and the distributed ledger object. These definitions are general and do not rely on the properties of the underlying distributed system, unless otherwise stated. In particular, they do not make any assumption on the types of failures that may occur in the processes (servers) that form the distributed system. However, we assume in this section that all the operations invoked by client processes eventually complete.

Distributed ledgers

A *distributed ledger object* (distributed ledger or DLO for short) is a concurrent ledger object that is implemented in a distributed manner. In particular, the ledger object is *implemented* by (and possibly replicated among) a set of (possibly distinct and geographically dispersed) computing devices that we refer to as *servers*. We refer to the processes that invoke the get() and append() operations of the

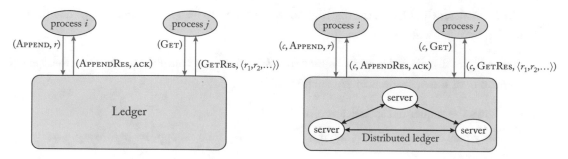

FIGURE 4.1: The interaction between processes and the ledger, where r, r_1, r_2, . . . are records. Left: General abstraction; right: Distributed ledger implemented by servers.

distributed ledger as *clients*. Figure 4.1 (right) depicts the interaction between the clients and the distributed ledger, implemented by servers.

In general, servers can fail. This leads to introducing mechanisms to achieve fault tolerance in the algorithm that implements the distributed ledger, like replicating the ledger. We assume in this section that these mechanisms guarantee that every operation invocation by a client will eventually have a matching response, and hence the operation is eventually completed. Additionally, the interaction of the clients with the servers may have to take into account the faulty nature of individual servers, as we discuss later in the next subsection.

Consistency of distributed ledgers

Distribution and replication intend to ensure availability and survivability of the ledger, in case a subset of the servers fails. At the same time, they raise the challenge of maintaining *consistency* among the different views that different clients get of the distributed ledger: what is the latest value of the ledger when multiple clients may send operation requests at different servers concurrently? Consistency semantics need to be in place to precisely describe the allowed values that a get() operation may return when it is executed concurrently with other get() or append() operations. Here, as examples, we provide the properties that operations must satisfy in order to guarantee *atomic consistency* or *linearizability* [Herlihy and Wing 1990], *sequential consistency* [Lamport 1979], and *eventual consistency* [Gentz and Dude 2017] semantics. In a similar way, other consistency guarantees, such as session consistency, could be formally defined [Gentz and Dude 2017].

Linearizability [Attiya and Welch 1994, Herlihy and Wing 1990] provides the illusion that the distributed ledger is accessed sequentially respecting the real time order, even when operations are invoked concurrently. That is, the distributed ledger seems to be a centralized ledger like the one implemented by Algorithm 4.2. Formally,[1]

Definition 4.8   A distributed ledger $\mathcal{L}$ is *linearizable* if, given any complete history $H_{\mathcal{L}}$, there exists a permutation σ of the operations in $H_{\mathcal{L}}$ such that:

1. σ follows the sequential specification of $\mathcal{L}$, and

2. for every pair of operations π_1, π_2, if $\pi_1 \to \pi_2$ in $H_{\mathcal{L}}$, then π_1 appears before π_2 in σ.

Sequential consistency [Attiya and Welch 1994, Lamport 1979] is weaker than linearizability in the sense that it only requires that operations respect the local ordering at each process, not the real time ordering. Formally,

Definition 4.9   A distributed ledger $\mathcal{L}$ is *sequentially consistent* if, given any complete history $H_{\mathcal{L}}$, there exists a permutation σ of the operations in $H_{\mathcal{L}}$ such that:

1. σ follows the sequential specification of $\mathcal{L}$, and

2. for every pair of operations π_1, π_2 invoked by a process p, if $\pi_1 \to \pi_2$ in $H_{\mathcal{L}}$, then π_1 appears before π_2 in σ.

Observe that in this context, *causal consistency* [Ahamad et al. 1995] is equivalent to sequential consistency. Both force a consistent permutation σ to respect the program order (property 2 in Definition 4.9). In addition, the causal order forces a causally consistent permutation σ to follow the sequential specification, because each get operation returns a sequence that exposes all the preceding append operations.

Let us finally give a definition of eventually consistent distributed ledgers. Informally speaking, a distributed ledger is eventually consistent if, for every append(r) operation that completes, *eventually* all get() operations return sequences that contain record r, and in the same position. Formally,

Definition 4.10   A distributed ledger $\mathcal{L}$ is *eventually consistent* if, given any complete history $H_{\mathcal{L}}$, there exists a permutation σ of the operations in $H_{\mathcal{L}}$ such that:

1. σ follows the sequential specification of $\mathcal{L}$, and

2. there exists a complete history $H'_{\mathcal{L}}$ that extends[2] $H_{\mathcal{L}}$ such that, for every complete history $H''_{\mathcal{L}}$ that extends $H'_{\mathcal{L}}$, every complete operation $\mathcal{L}$.get() in $H''_{\mathcal{L}} \setminus H'_{\mathcal{L}}$ returns a sequence that contains r, for all $\mathcal{L}$.append(r) $\in H_{\mathcal{L}}$.

---

[1] The formal definitions of linearizability and sequential consistency are adapted from Attiya and Welch [1994].
[2] A sequence X extends a sequence Y when Y is a prefix of X.

Observe that in the above definitions we consider $H_{\mathcal{L}}$ to be complete. As argued in Raynal [2013], the definitions can be extended to sequences that are not complete by reducing the problem of determining whether a complete sequence extracted by the non-complete one is consistent. That is, given a partial history $H_{\mathcal{L}}$, if $H_{\mathcal{L}}$ can be modified in such a way that every invocation of a non-complete operation is either removed or completed with a response event, and the resulting, complete sequence $H'_{\mathcal{L}}$ checks for consistency, then $H_{\mathcal{L}}$ also checks for consistency.

4.3.3 LINEARIZABLE DISTRIBUTED LEDGER IMPLEMENTATION IN A SYSTEM WITH CRASH FAILURES

To illustrate the viability and utility of the distributed ledger formulation, in this section we provide an implementation of a linearizable distributed ledger in an asynchronous distributed system with crash failures. Implementations of DLOs with the other weaker consistency guarantees can be found in Fernández Anta et al. [2018], and of a Byzantine-tolerant linearizable DLO in Cholvi et al. [2020].

Distributed setting and atomic broadcast

Consider an asynchronous message-passing distributed system. There is an unbounded number of clients accessing the distributed ledger. There is a set $\mathcal{S}$ of n servers that emulate a ledger (cf. Algorithm 4.2) in a distributed manner. Both clients and servers might fail by crashing. However, no more than $f < n$ servers might crash. Processes (clients and servers) interact by message-passing communication over asynchronous reliable channels.

Note that the sequential specification of a ledger $\mathcal{L}$ (cf. Definition 4.7) requires that two clients issuing two get operations will return two record sequences $\mathcal{L}.S$ and $\mathcal{L}.S'$ such that either $\mathcal{L}.S$ is a prefix of $\mathcal{L}.S'$ or vice versa. This property is called the *strong prefix* in Anceaume et al. [2019b] (cf. Section 4.4) and it essentially prevents forks—that is, having more than one record sequence at any given time. In both Anceaume et al. [2019b] and Fernández Anta et al. [2018], it was shown that consensus is required to implement such a property in a distributed setting.

For this purpose, as a building block, the implementation uses an atomic broadcast service (which in turn is built on consensus). Consequently, the correctness of the implementation depends on the modeling assumptions of the specific atomic broadcast service used. Specifically, the atomic broadcast service detailed in Défago et al. [2004] is considered, which has two operations: ABroadcast(m) used by a server to broadcast a message m to all servers $s \in \mathcal{S}$, and ADeliver(m) used by the atomic broadcast service to deliver a message m to a server. The service guarantees the total ordering of messages when up to $f < n/2$ may crash.

Algorithm 4.3: **External interface of a distributed ledger $\mathcal{L}$ executed by a client p**

---

1   $c \leftarrow 0$

2   **let** $L \subseteq \mathcal{S} : |L| \geq f + 1$

3   **function** $\mathcal{L}$.get( )

4      $c \leftarrow c + 1$

5   **send** request (c, GET) **to** the servers in L

6   **wait** response (c, GETRES, V) **from** some $i \in L$

7   **return** V

8   **function** $\mathcal{L}$.append(r)

9      $c \leftarrow c + 1$

10   **send** request (c, APPEND, r) **to** the servers in L

11   **wait** response $(c, \text{APPENDRES}, res)$ **from** some $i \in L$

12   **return** res

Client code

We assume that clients are aware of the faulty nature of servers and know (an upper bound on) the maximum number of faulty servers f. Hence, we assume they use a modified version of the interface presented in Algorithm 4.1 to deal with server unreliability. The new interface is presented in Algorithm 4.3. Every operation request is sent to a set L of at least $f + 1$ servers, to guarantee that at least one correct server receives and processes the request (if an upper bound on f is not known, then the clients contact all servers). Moreover, at least one such correct server will send a response that guarantees the termination of the operations. For formalization purposes, the first response received for an operation will be considered as the response event of the operation. In order to differentiate the different responses, all operations (and their requests and responses) are uniquely numbered with counter c, so duplicated responses will be identified and ignored (i.e., only the first one will be processed by the client).

Server code

We now focus on the code run by the servers. The servers take into account Algorithm 4.3, and in particular the fact that clients send the same request to multiple servers. This is important, for instance, to make sure that the same record r is not included in the sequence of records of the ledger multiple times. Algorithm 4.4 shows the pseudo-code used at the servers to implement a linearizable distributed ledger. The algorithm ensures that a record r appended by an append(r) operation is received by any succeeding get() operation, even if the two operations were invoked at

---

Algorithm 4.4: Linearizable distributed ledger; code for server i

1   **Init:** $S_i \leftarrow \emptyset$; $get\_pending_i \leftarrow \emptyset$; $pending_i \leftarrow \emptyset$

2   **receive** (c, GET) **from** process p

3   ABroadcast(get, p, c)

4   add (p, c) to $get\_pending_i$

5   **end receive**

6   **upon** ADeliver(get, p, c) **do**

7   **if** $(p, c) \in get\_pending_i$ **then**

8   **send** response (c, GETRES, S_i) **to** p

9   remove (p, c) from $get\_pending_i$

10   **end upon**

11   **receive** (c, APPEND, r) **from** process p

12   ABroadcast$(append, r)$

13   add (c, r) to $pending_i$

14   **end receive**

15   **upon** ADeliver$(append, r)$ **do**

16   **if** $r \notin S_i$ **then**

17       $S_i \leftarrow S_i \parallel r$

18   **if** $\exists (c, r) \in pending_i$ **then**

19   **send** response $(c, \text{APPENDRES}, \text{ACK})$ **to** $r.p$

20   remove (c, r) from $pending_i$

21   **end upon**

---

different clients. This algorithm resembles approaches used in Wang et al. [2014] for implementing arbitrary objects, and Attiya and Welch [2004], Chaudhuri et al. [1993], and Mavronicolas and Roth [1999] for implementing consistent read/write objects.

Briefly, when a server receives a get or an append request, it adds the request in a pending set and atomically broadcasts the request to all other servers. When an append or get message is delivered, the server then replies to the requesting process (if it did not reply yet). This leads to the following result.

Theorem 4.11 ([Fernández Anta et al. 2018 (Theorem 2)])   *The combination of Algorithm 4.3 and Algorithm 4.4 implements a linearizable distributed ledger.*

Algorithm 4.5: Validated ledger $\mathcal{V}$ (centralized)

---

1 **Init:** $S \leftarrow \emptyset$

2 **receive** (GET) **from** process p
3   **send** response (GETRES, S) **to** p
4 **end receive**

5 **receive** (APPEND, r) **from** process p
6   **if** $Valid(S \parallel r)$ **then**
7          $S \leftarrow S \parallel r$
8   **send** response (APPENDRES, ACK) **to** p
9   **else send** response (APPENDRES, NACK) **to** p
10 **end receive**

4.3.4   VALIDATED LEDGERS

A *validated ledger* $\mathcal{V}$ is one in which specific semantics are imposed on the contents of the records stored in it. For instance, if the records are (Bitcoin-like) financial transactions, the ledger's semantics should prevent problems such as double spending, or apply other transaction validation used as part of the Bitcoin protocol [Nakamoto 2008]. The ledger preserves the semantics with a validity check in the form of a Boolean function *Valid*() that takes as an input a sequence of records S and returns *true* if and only if the semantics are preserved. In a validated ledger the result of an $\text{append}_p(r)$ operation may be NACK if the validity check fails. Algorithm 4.5 presents a centralized implementation of a validated ledger $\mathcal{V}$. The *Valid*() function is similar to the one used to check validity in Crain et al. [2017] or the external validity in Cachin et al. [2001], but these are used in the consensus algorithm to prevent it from deciding on an invalid value. In Algorithm 4.5 an invalid record is locally detected and discarded.

The sequential specification of a validated ledger must take into account the possibility that an append returns NACK. Accordingly, property (2) of Definition 4.7 must be revised as follows:

Definition 4.12   The sequential specification of a **validated** ledger $\mathcal{V}$ over the sequential history $H_{\mathcal{V}}$ is defined as follows. The value of the sequence $\mathcal{V}.S$ is initially the empty sequence. If at the invocation event of an operation π in $H_{\mathcal{V}}$ the value of the sequence in ledger $\mathcal{V}$ is $\mathcal{V}.S = V$, then:

1. If π is a $\mathcal{V}.\text{get}_p()$ operation, then the response event of π returns V.
2.   (a) If π is a $\mathcal{V}.\text{append}_p(r)$ operation that returns ACK, then $Valid(V \parallel r) = true$ and at the response event of π, the value of the sequence in ledger $\mathcal{V}$ is $\mathcal{V}.S = V \parallel r$.

(b) If π is a $\mathcal{V}$.append$_p(r)$ operation that returns NACK, then $Valid(V \parallel r) = false$ and at the response event of π, the value of the sequence in ledger $\mathcal{V}$ is $\mathcal{V}.S = V$.

Based on this revised notion of sequential specification, one can define the eventual, sequential, causal, and linearizable consistent validated distributed ledger in a similar manner as in Section 4.3.2.

Regarding the implementation of validated distributed ledgers, these can be derived by extending distributed ledgers' implementations to provide validity. For example, in Algorithm 4.4, in the append operation, in Line 15, besides checking whether the record is already in the ledger, its validity could also be checked (call $Valid()$ as in Line 6 of Algorithm 4.5). In the case where it is not found valid, the server would not append the record in the ledger, and instead it would return NACK to the client.

4.4   BLOCKCHAIN ABSTRACT DATA TYPE

In Section 4.3 we focused on *permissioned blockchains*, and constructed DLO abstract data types with a totally ordered sequence of records. In Section 4.4, we present another abstract data type, the BlockTree, that allows us to deal with *permissionless blockchains*. To account for forks, this abstract data type is a *tree* of records.

In particular, we are interested in formalizing the blockchains as a family of formal specifications that can be combined to meet specific consistency criteria. To this end the authors in Anceaume et al. [2019b] provide specifications as a composition of *abstract data types* (whose definition can be recalled from Section 4.2) together with a hierarchy of *consistency criteria* that formally characterizes the histories admissible for distributed programs that use them. The advantage of specifying shared objects as abstract data types over implementation-based alternatives such as Girault et al. [2018] is the possibility to reason on the consistency of a system independently of the communication model [Perrin et al. 2016]. More precisely, the authors in Anceaume et al. [2019b] define two abstract data types, the *BlockTree* and the *oracle*. The BlockTree models the blockchain data structure, which is a tree of blocks providing an append and a read operation. The append operation is intended to insert a new leaf in the tree, provided that the newly inserted block is valid. The validity property of a block in the BlockTree is abstracted as a general predicate, which is application dependent. For instance, in Bitcoin a valid block is a block containing a hash hs of three fields: a set of non–double spending transactions, the hash of the previous block, and a nonce such that hs starts with a given number of leading zeros.

The *oracle* abstracts the mechanism used to grant to processes the insertion of new blocks in the blockchain, such as proof of work or other agreement mechanisms. To this end the oracle is modeled as a simple token manager. A token is generated when a process obtains a valid block, and it is consumed when the process obtains the right to insert the block into the chain. Those notions are presented in Sections 4.4.1 and 4.4.2.

Modelization of the blockchain with two different abstractions has several benefits. The first advantage is the possibility of dealing with liveness properties at finer granularity level—separating the termination of the validation process, managed by the oracle, from the termination of updating a replicated data structure, managed by the BlockTree. That separation is very useful because in many implementations the proof-of-* mechanisms (proof-of-work, proof-of-elapsed-time, proof-of-space, etc.) are a local process, while the process of updating a replicated data structure is a global computation.

Furthermore, the possibility to reason on the BlockTree in isolation allows us to extend the consistency criteria theory with a new consistency criterion for BlockTree that captures the eventual convergence process in blockchain systems. This criterion, presented in Section 4.4.1, is a weaker consistency criterion than the so-called monotonic prefix consistency criterion introduced in Girault et al. [2018]. The monotonic prefix consistency criterion informally says that any two reads return two chains such that one is the prefix of the other. A relaxation of this consistency criterion, *eventual prefix*, defined in Anceaume et al. [2019b], admits any two chains having a divergent prefix for a finite interval of the history. The *strong prefix* property is further introduced in Anceaume et al. [2019b]. This consistency criterion is equivalent, under certain conditions, to the monotonic prefix consistency criterion. The authors of Anceaume et al. [2019b] define two BlockTree abstracts, one verifying the eventual prefix property and the other one verifying the strong prefix property.

The oracle proposed in Anceaume et al. [2019b] has two different versions: *prodigal* "without memory" and *frugal* "with memory." The prodigal oracle does not remember how many tokens are consumed for a given block to extend, while the frugal oracle counts the number of tokens actually consumed for any block. More specifically, the prodigal oracle does not control the number of forks in the system, while the frugal oracle restricts the number of forks to k.

In Anceaume et al. [2019b] the authors prove the necessary conditions that must be verified by the communication model to implement the BlockTree with the eventual prefix in a message-passing system. Interestingly, the authors show the necessity of a light form of reliable broadcast enjoying both the Validity property (if a correct process sends a message then it eventually delivers it) and the Agreement property (if a message is delivered by some correct process then the message is eventually delivered by every correct process).

Section 4.4.5 maps specifications to existing representative blockchains. It is demonstrated in Anceaume et al. [2019b] that Bitcoin and Ethereum implement a prodigal oracle and their executions are eventually consistent, while Algorand implements a frugal oracle and its executions are strongly consistent with high probability, and ByzCoin, PeerCensus, Red Belly, and Hyperledger executions (also implementing a frugal oracle) are strongly consistent.

In what follows, we introduce the BlockTree and the token oracle ADTs along with their consistency criteria. All the notations, definitions, and results are borrowed from Anceaume et al. [2019b].

4.4.1   BLOCKTREE ADT

The data structure implemented by blockchain-like systems is formalized in Anceaume et al. [2019b] as a *directed rooted tree* $bt = (V_{bt}, E_{bt})$ called *BlockTree*. Each vertex of the BlockTree is a *block* and any edge points backward to the root, called the *genesis block*. The height of a block refers to its distance to the root and b_k denotes a block located at height k. By convention, the root of the BlockTree is denoted by b_0. Blocks are said to be valid if they satisfy a predicate P that is application dependent (for instance, in Bitcoin, a block is considered valid if it can be connected to the current blockchain and does not contain transactions that double spend a previous transaction). Let $\mathcal{B}$ be a countable and non-empty set of blocks and let $\mathcal{B}' \subseteq \mathcal{B}$ be a countable and non-empty set of valid blocks (i.e., $\forall b \in \mathcal{B}'$, $P(b) = \top$). By assumption $b_0 \in \mathcal{B}'$; let us denote by $\mathcal{BC}$ a countable non-empty set of blockchains, where a blockchain is a path from a leaf of bt to b_0. A blockchain is denoted by bc. Let $\mathcal{F}$ be a countable non-empty set of selection functions, $f \in \mathcal{F} : \mathcal{BT} \to \mathcal{BC}$; $f(bt)$ selects a blockchain bc from the BlockTree bt (note that b_0 is not returned). Selection function f is a parameter of the ADT, which is encoded in the state and does not change over the computation. This reflects, for instance, the longest chain or the heaviest chain used in some blockchain implementations.

The following notations are also used: $\{b_0\}^\frown f(bt)$ represents the concatenation of b_0 with the blockchain of bt; and $\{b_0\}^\frown f(bt)^\frown\{b\}$ represents the concatenation of b_0 with the blockchain of bt and a block b.

Sequential specification of the BlockTree
The sequential specification of the BlockTree is defined as follows.

Definition 4.13 (BlockTree ADT (BT-ADT))   The BlockTree abstract data type is the 6-tuple BT-ADT$=\langle A = \{\text{append}(b), \text{read}() : b \in \mathcal{B}\}, B = \mathcal{BC} \cup \{\text{true, false}\}, Z = \mathcal{BT} \times \mathcal{F}, \xi_0 = (bt^0, f), \tau, \delta\rangle$, where the transition function $\tau : Z \times A \to Z$ is defined by

- $\tau((bt, f), \text{append}(b)) = (\{b_0\}^\frown f(bt)^\frown\{b\}, f)$ if $b \in \mathcal{B}'$; (bt, f) otherwise;
- $\tau((bt, f), \text{read}()) = (bt, f)$,

and the output function $\delta : Z \times A \to B$ is defined by

- $\delta((bt, f), \text{append}(b)) = \text{true}$ if $b \in \mathcal{B}'$; false otherwise;
- $\delta((bt, f), \text{read}()) = \{b_0\}^\frown f(bt)$;
- $\delta((bt_0, f), \text{read}()) = b_0$.

The semantics of the read and the append operations directly depend on the selection function $f \in \mathcal{F}$. In Anceaume et al. [2019b] the authors let this function remain generic to capture different blockchain implementations. In the same way, predicate P is kept unspecified. The predicate P mainly abstracts the creation process of a block, which may fail or successfully terminate. This process will be further specified in Section 4.4.2.

Concurrent specification of a BT-ADT and consistency criteria

The concurrent specification of the BT-ADT is a set of concurrent histories. A *BT-ADT* consistency criterion is a function that returns the set of concurrent histories admissible for a BlockTree abstract data type. In Anceaume et al. [2019b] the authors define two *BT* consistency criteria: *BT strong consistency* and *BT eventual consistency*. For ease of readability, the following notations are used:

- $E(a^*, r^*)$ is an infinite set containing an infinite number of append() and read() invocation and response events.

- $E(a, r^*)$ is an infinite set containing (i) a finite number of append() invocation and response events and (ii) an infinite number of read() invocation and response events.

- $e_{inv}(o)$ and $e_{rsp}(o)$ indicate respectively the invocation and response events of an operation o, and $e_{rsp}(r) : bc$ denotes the returned blockchain bc associated with the response event $e_{rsp}(r)$.

- score : $\mathcal{BC} \to \mathbb{N}$ denotes a monotonic increasing deterministic function that takes as input a blockchain bc and returns a natural number s as the score of bc, which can be the height, the weight, etc. Informally the authors refer to such a value as the score of a blockchain; by convention the authors refer to the score of the blockchain uniquely composed by the genesis block as s_0 (i.e., score($\{b_0\}) = s_0$). Increasing monotonicity means that score($bc^\frown\{b\}$) > score(bc).

- mcps : $\mathcal{BC} \times \mathcal{BC} \to \mathbb{N}$ is a function that, given two blockchains bc and bc', returns the score of the maximal common prefix between bc and bc'.

- $bc \sqsubseteq bc'$ iff bc prefixes bc'.

BT strong consistency

The BT strong consistency criterion is the conjunction of the following four properties. The block validity property imposes the requirement that each block in a blockchain returned by a read() operation is *valid* (i.e., satisfies predicate P) and has been inserted in the BlockTree with the append() operation. The Local monotonic read property states that, given the sequence of read() operations at the same process, the score of the returned blockchain never decreases. The strong prefix property states that for each couple of read operations, one of the returned blockchains is a prefix of the other returned one (i.e., the prefix never diverges). Finally, the Ever-growing tree states that scores of returned blockchains eventually grow. More precisely, let s be the score of the blockchain returned by a read() response event r in $E(a^*, r^*)$; then for each read() operation r, the set of read() operations such that $e_{rsp}(r) \nearrow e_{inv}(r')$ that do not return blockchains with a score greater than s is finite. More formally, the BT strong consistency criterion is defined as follows:

Definition 4.14 (BT strong consistency criterion (SC)) A concurrent history $H = \langle \Sigma, E, \Lambda, \mapsto, \prec, \nearrow \rangle$ of the system that uses a BT-ADT verifies the BT strong consistency criterion if the following properties hold:

Block validity. $\forall e_{rsp}(r) \in E, \forall b \in e_{rsp}(r) : bc, b \in \mathcal{B}' \wedge \exists e_{inv}(\mathsf{append}(b)) \in E.$

Local monotonic read. $\forall e_{rsp}(r), e_{rsp}(r') \in E^2$, if $e_{rsp}(r) \mapsto e_{inv}(r')$, then $\mathsf{score}(e_{rsp}(r) : bc) \leq \mathsf{score}(e_{rsp}(r') : bc').$

Strong prefix. $\forall e_{rsp}(r), e_{rsp}(r') \in E^2$, $(e_{rsp}(r) : bc \sqsubseteq e_{rsp}(r) : bc') \vee (e_{rsp}(r) : bc \sqsubseteq e_{rsp}(r') : bc').$

Ever-growing tree. $\forall e_{rsp}(r) \in E(a^*, r^*), s = \mathsf{score}(e_{rsp}(r) : bc)$, then $|\{e_{inv}(r') \in E \mid e_{rsp}(r) \nearrow e_{inv}(r'), \mathsf{score}(e_{rsp}(r') : bc) \leq s\}| < \infty.$

BT eventual consistency

The BT eventual consistency criterion is the conjunction of the block validity, the Local monotonic read, and the Ever-growing tree of the BT strong consistency criterion together with the eventual prefix, which states that for each blockchain returned by a read() operation with s as score, eventually all the read() operations will return blockchains sharing the same maximum common prefix at least up to s. To say it differently, let H be a history with an infinite number of read() operations, and let s be the score of the blockchain returned by a read() r; then the set of read() operations r', such that $e_{rsp}(r) \nearrow e_{inv}(r')$, that do not return blockchains sharing the same prefix at least up to s is finite.

Definition 4.15 (Eventual prefix property [Anceaume et al. 2019b]) Given a concurrent history $H = \langle \Sigma, E(a, r^*), \Lambda, \mapsto, \prec, \nearrow \rangle$ of the system that uses a BT-ADT, the authors denote by s, for any read operation $r \in \Sigma$ such that $\exists e \in E(a, r^*), \Lambda(r) = e$, the score of the returned blockchain—that is, $s = \mathsf{score}(e_{rsp}(r) : bc)$. In Anceaume et al. [2019b], the authors denote by E_r the set of response events of read() operations that occurred after r response—that is, $E_r = \{e \in E \mid \exists r' \in \Sigma, r' = \mathsf{read}, e = e_{rsp}(r') \wedge e_{rsp}(r) \nearrow e_{rsp}(r')\}$. Then, H satisfies the eventual prefix property if for all read() operations $r \in \Sigma$ with score s,

$$|\{(e_{rsp}(r_h), e_{rsp}(r_k)) \in E_r^2 | h \neq k, \mathsf{mpcs}(e_{rsp}(r_h) : bc_h, e_{rsp}(r_k) : bc_k) < s\}| < \infty.$$

The eventual prefix properties capture the fact that two or more concurrent blockchains can co-exist in a finite interval of time, but that all the participants adopt the same branch for each cut of the history. This cut of the history is defined by a read() that picks up a blockchain with a given score.

Based on this definition, the BT eventual consistency criterion is defined as follows:

Definition 4.16 (BT eventual consistency criterion $\Diamond C$) A concurrent history $H = \langle \Sigma, E, \Lambda, \mapsto, \prec, \nearrow \rangle$ of the system that uses a BT-ADT verifies the *BT eventual consistency criterion* if it satisfies the block validity, Local monotonic read, Ever-growing tree, and the eventual prefix properties.

Let us remark that the BlockTree allows at any time for creating a new branch in the tree, which is called a *fork* in the blockchain literature. Moreover, an append is successful only if the input block is valid with respect to a predicate. This means that histories with no append() operations are trivially admitted.

We next introduce a new abstract data type called a token oracle that, when combined with the BlockTree, will help in (i) validating blocks and (ii) controlling forks.

4.4.2   TOKEN ORACLE Θ-ADT

In this section we present the formalization of the token oracle Θ introduced in Anceaume et al. [2019b] to capture the creation of blocks in the BlockTree structure. The block creation process requires that the new block must be closely related to an already existing valid block in the BlockTree structure. We abstract this implementation-dependent process by assuming that a process will obtain the right to chain a new block b_ℓ to b_h if it successfully gains a token tkn_h from the token oracle Θ. Once obtained, the proposed block b_ℓ is considered as valid, and will be denoted by $b_\ell^{tkn_h}$. By construction $b_\ell^{tkn_h} \in \mathcal{B}'$. In the following, in order to be as general as possible, blocks are modeled as objects. More formally, when a process wants to access a generic object obj_h, it invokes the getToken(obj_h, obj_ℓ) operation with object obj_ℓ from set $\mathcal{O} = \{obj_1, obj_2, \ldots\}$. If the getToken($obj_h, obj_\ell$) operation is successful, it returns an object $obj_\ell^{tkn_h} \in \mathcal{O}'$, where (i) tkn_h is the token required to access object obj_h and (ii) each object $obj_k \in \mathcal{O}'$ is valid with respect to predicate P; that is, $P(obj_k) = \top$. A token is *generated* each time it is provided to a process and it is *consumed* when the oracle grants the right to connect it to the previous object. Each token can be consumed at most once. To consume a token we define the token consumption as the consumeToken($obj_\ell^{tkn_h}$) operation, where the consumed token tkn_h is the token required for the object obj_h. A maximal number of tokens k for an object obj_h is managed by the oracle. The consumeToken($obj_\ell^{tkn_h}$) operation's side-effect on the state is to decrement k by one for object obj_h.

Below, we specify two token oracles that differ in the way tokens are managed. The first oracle, called *prodigal* and denoted by Θ_P, has no upper bound on the number of tokens consumed for an object, while the second oracle, Θ_F, is called *frugal* and ensures that no more than k tokens can be consumed for each object.

Oracle Θ_P, when combined with the BlockTree abstract data type, will only help in validating blocks, while oracle Θ_F manages tokens in a more controlled way to guarantee that no more than k forks can occur on a given block.

Θ_P-ADT and Θ_F-ADT definitions

For both oracles, when the getToken(obj_k, obj_h) operation is invoked, the oracle provides a token with a certain probability p_{α_i} where α_i is a "merit" parameter characterizing the invoking process i.[3] Note that the oracle knows α_i of the invoking process i, which might be unknown to the process itself. For each merit α_i, the state of the token oracle embeds an infinite tape where each cell of the tape contains either tkn or $\bot$. Since each tape is identified by a specific α_i and p_{α_i}, we assume that each tape contains a pseudorandom sequence of values in $\{tkn, \bot\}$ depending on α_i.[4] When a getToken(obj_k, obj_h) operation is invoked by a process with merit α_i, the oracle pops the first cell from the tape associated to α_i, and a token is provided to the process if that cell contains tkn.

Both oracles also enjoy an infinite array of counters, one for each object; this array is decreased each time a token is consumed for a specific object. When the counter reaches 0, then no more tokens can be consumed for that object. For the sake of generality, Θ_P is defined as Θ_F with $k = \infty$ while for Θ_F each counter is initialized to $k \in \mathbb{N}$.

The following definitions and notations are borrowed from Anceaume et al. [2019b].

- $\mathcal{O} = \{obj_1, obj_2, \ldots\}$, an infinite set of generic objects uniquely identified by their index i.

- $\mathcal{O}' \subset \mathcal{O}$, the subset of objects valid with respect to predicate P, i.e., $\forall obj_i' \in \mathcal{O}'$, $P(obj_i') = \top$.

- $\mathfrak{T} = \{tkn_1, tkn_2, \ldots\}$, an infinite set of tokens.

- $\mathcal{A} = \{\alpha_1, \alpha_2, \ldots\}$, an infinite set of rational values.

- $\mathcal{M}$ is a countable non-empty set of mapping functions $\mathsf{m}(\alpha_i)$ that generate an infinite pseudorandom tape, $tape_{\alpha_i}$, such that the probability to have the string tkn in a cell is related to a specific α_i, $\mathsf{m} \in \mathcal{M} : \mathcal{A} \to \{tkn, \bot\}^*$.

- $K[\,]$ is an infinite array of counters (one per object). All the counters are initialized with a $k \in \mathbb{N}$, where k is a parameter of the oracle ADT.

- pop: $\{tkn, \bot\}^* \to \{tkn, \bot\}^*$, pop($a \cdot w$) $= w$.

- head: $\{tkn, \bot\}^* \to \{tkn, \bot\}^*$, head($a \cdot w$) $= a$.

- dec: $\{K\} \times \mathbb{N} \to \{K\}$, dec($K, i$) $= K : K[i] = K[i] - 1$ **if** $K[i] > 0$; $K[i] = 0$ otherwise.

- get: $\{K\} \times \mathbb{N} \to \mathbb{N}$, get($K, i$) $= K[i]$.

---

[3] The merit parameter can reflect, for instance, the hashing rate of the invoking process.

[4] It is assumed to be a pseudorandom sequence mostly indistinguishable from a Bernoulli sequence consisting of a finite or infinite number of independent random variables $X1, X2, X3, \ldots$ such that (i) for each k, the value of X_k is either tkn or $\bot$; and (ii) $\forall X_k$ the probability that $X_k = tkn$ is p.

Definition 4.17 (Θ_F-ADT)   The Θ_F abstract data type is the 6-tuple Θ_F-ADT = (A = {getToken (obj_h, obj_ℓ), consumeToken$(obj_\ell^{tkn_h})$: obj_h, $obj_\ell^{tkn_h} \in \mathcal{O}'$, $obj_\ell \in \mathcal{O}$, $tkn_h \in \mathfrak{T}$}, B = $\mathcal{O}' \cup$ *Boolean*, Z = $m(A)^* \times \{K\} \cup \{$pop, head, dec, get$\}$, $\xi_0, \tau, \delta)$, where the transition function $\tau : Z \times A \to Z$ is defined by

- $\tau((\{tape_{\alpha_1}, \ldots, tape_{\alpha_i}, \ldots\}, K)$, getToken$(obj_h, obj_\ell))$
  $= (\{tape_{\alpha_1}, \ldots, \text{pop}(tape_{\alpha_i}), \ldots\}, K)$ with α_i the merit of the invoking process.

- $\tau((\{tape_{\alpha_1}, \ldots, tape_{\alpha_i}, \ldots\}, K)$, consumeToken$(obj_\ell^{tkn_h}))$
  $= (\{tape_{\alpha_1}, \ldots, tape_{\alpha_i}, \ldots\}, \text{dec}(K, h))$,   if $tkn_h \in \mathfrak{T}$; $\{(\{tape_{\alpha_1}, \ldots, tape_{\alpha_i}, \ldots\}, K)\}$ otherwise.

The output function $\delta : Z \times A \to B$ is defined by

- $\delta((\{tape_{\alpha_1}, \ldots, tape_{\alpha_i}, \ldots\}, K)$, getToken$(obj_h, obj_\ell)) = obj_\ell^{tkn_h} : obj_\ell^{tkn_h} \in \mathcal{O}'$, $tkn_h \in \mathfrak{T}$, if head$(tape_{\alpha_i}) = tkn$ with α_i the merit of the invoking process; $\perp$ otherwise.

- $\delta((\{tape_{\alpha_1}, \ldots, tape_{\alpha_i}, \ldots\}, K)$, consumeToken$(obj_\ell^{tkn_h})) = \top$ if $tkn_h \in \mathfrak{T}$ and get$(K, h) > 0$; $\perp$ otherwise.

Definition 4.18 (Θ_P-ADT)   The Θ_P abstract data type is defined as the Θ_F-ADT with $k = \infty$.

4.4.3   BT-ADT AUGMENTED WITH Θ ORACLES

In this section we augment the BT-ADT with Θ oracles and then analyze the histories generated by their combination. Specifically, following the work in Anceaume et al. [2019b], we define a refinement of the append(b_ℓ) operation of the BT-ADT with the oracle operations. A generic implementation (cf. Algorithm 4.6) of the BT-ADT invokes the getToken$(b_k \leftarrow$ last_block$(f(bt))$, $b_\ell)$ operation as long as it returns a token on b_k—that is, $b_\ell^{tkn_h}$, which is a valid block in $\mathcal{B}'$. Once obtained, the token is consumed and the append terminates; that is, the block $b_\ell^{tkn_h}$ is appended to the blockchain $f(bt)$. Notice that those two operations and the concatenation occur atomically.

   The BT-ADT augmented with the Θ_F or Θ_P oracle is a *refinement* $\mathfrak{R}(\text{BT-ADT}, \Theta_F)$ or $\mathfrak{R}(\text{BT-ADT}, \Theta_P)$ respectively.

Definition 4.19 ($\mathfrak{R}(\text{BT-ADT}, \Theta_F)$ refinement)   Given the abstract data type BT-ADT = $\langle A, B, Z, \xi_0, \tau, \delta \rangle$, and the Θ_F-ADT = $(A^\Theta, B^\Theta, Z^\Theta, \xi_0^\Theta, \tau^\Theta, \delta^\Theta)$, we have $\mathfrak{R}(\text{BT-ADT}, \Theta_F) = \langle A' = A \cup A^\Theta$, $B' = B \cup B^\Theta$, $Z' = Z \cup Z^\Theta$, $\xi_0' = \xi_0 \cup \xi_0^\Theta$, $\tau', \delta' \rangle$, where the transition function $\tau' : Z' \times A' \to Z'$ is defined by

Algorithm 4.6: A generic implementation fragment of the BT-ADT that employs a Θ oracle to implement the append() operation [Anceaume et al. 2019b]

1 **Init:**
2      $token \leftarrow \bot$;
3      $bt_i \leftarrow b_0$;
4   . . .

5 **upon** append(b_ℓ) **do**
6   . . .
7   **while** $token = \bot$ **do**
8          $token \leftarrow$ getToken($b_n \leftarrow$ last_block($f(bt)$), b_ℓ)
9   consumeToken($token$) $\wedge \{b_0\}^\frown f(bt)^\frown \{b\}$
10     $token \leftarrow \bot$
11   . . .
12 **end upon**

$$\tau_a = \tau'((\{tape_{\alpha_1}, \ldots, tape_{\alpha_i}, \ldots\}, K, bt, f), \text{getToken}(b_k \leftarrow \text{last\_block}(bt), b_\ell))$$

$$= (\{tape_{\alpha_1}, \ldots, \text{pop}(tape_{\alpha_i}), \ldots\}, K, bt, f).$$

$$\tau_b = \tau'((\{tape_{\alpha_1}, \ldots, tape_{\alpha_i}, \ldots\}, K, bt, f), \text{consumeToken}(b_\ell^{tkn_h}))$$

$$= (\{tape_{\alpha_1}, \ldots, tape_{\alpha_i}, \ldots\}, \text{dec}(K, h), \{b_0\}^\frown f(bt)^\frown \{b\}, f) \quad \text{if } tkn_h \in \mathfrak{T}$$

$$= (\{tape_{\alpha_1}, \ldots, tape_{\alpha_i}, \ldots\}, K, bt, f) \quad \text{otherwise.}$$

$$\tau'((\{tape_{\alpha_1}, \ldots, tape_{\alpha_i}, \ldots\}, K, bt, f), \text{append}(b)) = \tau_b \circ \tau_a^*.$$

$$\tau'(\{tape_{\alpha_1}, \ldots, tape_{\alpha_i}, \ldots\}, K, bt, f, \text{read}()) = bt,$$

where $\tau_b \circ \tau_a^*$ is the repeated application of τ_a until

$$\delta_a((\{tape_{\alpha_1}, \ldots, tape_{\alpha_i}, \ldots\}, K, bt, f), \text{getToken}(b_k \leftarrow \text{last\_block}(bt), b_\ell)) = b_\ell^{tkn_h}$$

concatenated with the τ_b application.

Additionally, the output function $\delta' : Z \times A \to B$ is defined by:

- $\delta_a = \delta'((\{tape_{\alpha_1}, \ldots, tape_{\alpha_i}, \ldots\}, K, bt, f), \text{getToken}(b_k \leftarrow \text{last\_block}(bt), b_\ell)) = b_\ell^{tkn_h}$: $b_\ell^{tkn_h} \in \mathcal{B}'$, $tkn_h \in \mathfrak{T}$, if head($tape_{\alpha_i}$) $= tkn$ with α_i the merit of the invoking process; $\bot$ otherwise.

- $\delta_b = \delta'((\{tape_{\alpha_1}, \ldots, tape_{\alpha_i}, \ldots\}, K, bt, f), \text{consumeToken}(obj_\ell^{;tkn_h})) = \top$ if $tkn_h \in \mathfrak{T}$ and $\text{get}(K, h) > 0$; $\bot$ otherwise.

- $\delta'((\{tape_{\alpha_1}, \ldots, tape_{\alpha_i}, \ldots\}, K, bt, f), \text{append}(b)) = \delta_b \circ \delta_a^*$.

- $\delta'((\{tape_{\alpha_1}, \ldots, tape_{\alpha_i}, \ldots\}, K, bt, f), \text{read}()) = \{b_0\}^\frown f(bt)$.

- $\delta'((\{tape_{\alpha_1}, \ldots, tape_{\alpha_i}, \ldots\}, K, bt_0, f), \text{read}()) = b_0$.

where $\delta_b \circ \delta_a^*$ is the repeated application of δ_a until

$$\delta_a((\{tape_{\alpha_1}, \ldots, tape_{\alpha_i}, \ldots\}, K, bt, f), \text{getToken}(\text{last\_block}(bt), b)) = b_\ell^{tkn_h}$$

concatenated with the δ_b application.

Definition 4.20 ($\mathfrak{R}(\text{BT-ADT}, \Theta_P)$ refinement)   Same definition as the $\mathfrak{R}(\text{BT-ADT}, \Theta_F)$ refinement.

Definition 4.21 (k-Fork Coherence)   A concurrent history $H = \langle \Sigma, E, \Lambda, \mapsto, \prec, \nearrow \rangle$ of the BT-ADT composed with Θ_F-ADT satisfies the *k-Fork Coherence* if there are at most k append() operations that return $\top$ for the same token.

Theorem 4.22 (k-Fork Coherence [Anceaume et al. 2019b])   *Each concurrent history $H = \langle \Sigma, E, \Lambda, \mapsto, \prec, \nearrow \rangle$ of the BT-ADT composed with a Θ_F-ADT satisfies the* k-Fork Coherence.

4.4.4   ON THE IMPLEMENTATION OF BT-ADTS

Implementation of BT-ADTs in shared memory
In Anceaume et al. [2018] the authors prove that $\Theta_{F,k=1}$ has consensus number ∞ and that Θ_P has consensus number 1. We consider a concurrent system composed of n processes such that up to f processes are faulty (stop prematurely by crashing), $f < n$. Moreover, processes can communicate through atomic registers.

Frugal with $k = 1$ at least as strong as Consensus
In Anceaume et al. [2018] the authors prove that there exists a wait-free implementation of Consensus [Lamport et al. 1982] by the $\Theta_{F,k=1}$ oracle object. In particular, in this case $\Theta_{F,k=1} = \langle A = \{\text{getToken}(b_h, b_\ell), \text{consumeToken}(b_\ell^{tkn_h}) : b_h, b_\ell^{tkn_h} \in \mathcal{B}', b_\ell \in \mathcal{B}, tkn_h \in \mathfrak{T}\}, B = \mathcal{B}' \cup Boolean, Z = m(\mathcal{A})^* \times \{K\} \times k \cup \{\text{pop}, \text{head}, \text{dec}, \text{get}\}, \xi_0, \tau, \delta \rangle$. We explicitly consider blocks and valid blocks ($\mathcal{B}$ and $\mathcal{B}'$) rather than objects and valid objects ($\mathcal{O}$ and $\mathcal{O}'$). Moreover, we consider a version of the definition of the Consensus problem tailored for blockchains. Thus, we consider the Validity property as in Crain et al. [2017] such that the decided block b satisfies some validity predicate P.

Definition 4.23   Consensus $\mathcal{C}$:

Termination.   Every correct process eventually decides some value.

Integrity.   No correct process decides twice.

Agreement.   If there is a correct process that decides a value b, then eventually all the correct processes decide b.

Validity [Crain et al. 2017].   A decided value is valid; it satisfies the predefined predicate denoted P.

In Anceaume et al. [2018] the authors first prove that there exists a wait-free implementation of the Compare&Swap() object by the consumeToken() object in the case of $\Theta_{F,k=1}$, implying that consumeToken() has the same Consensus number as Compare&Swap(), which is ∞ (see Herlihy [1991]). Finally, they compose the consumeToken() with the getToken() object, proving that there exists a wait-free implementation of Consensus $\mathcal{C}$ by $\Theta_{F,k=1}$.

Algorithms 4.7 and 4.8 respectively describe consumeToken() (CT), as specified by the Θ-ADT, and the Compare&Swap() (CAS). Compare&Swap() takes three parameters as input: the *register*, the *old_value*, and the *new_value*. If the value in *register* is the same as *old_value*, then the *new_value* is stored in *register* and in any case the operation returns the value that was in *register*

Algorithm 4.7: consumeToken() in the case of $\Theta_{F,k=1}$ [Anceaume et al. 2018]

```
1  function consumeToken(b_ℓ^{tkn_h})
2      previous_value ← K[h]
3      if previous_value = {} ∧ tkn_h ∈ 𝔗 then
4          K[h] ← K[h] ∪ {b_ℓ^{tkn_h}}
5      return K[h]
```

Algorithm 4.8: Compare&Swap() in the case of $\Theta_{F,k=1}$ [Anceaume et al. 2018]

```
1  function Compare&Swap(register, old_value, new_value)
2      previous_value ← register
3      if previous_value = old_value then
4          register ← new_value
5      return previous_value
```

Algorithm 4.9: An implementation of CAS by CT in the case of $\Theta_{F,k=1}$ [Anceaume et al. 2018]

1 **function** Compare&Swap($K[h], \{\}, b_\ell^{tkn_h}$)
2      $returned\_value \leftarrow$ consumeToken($b_\ell^{tkn_h}$)
3   **if** $returned\_value = b_\ell^{tkn_h}$ **then**
4   **return** $\{\}$
5   **else**
6   **return** $returned\_value$

Algorithm 4.10: The protocol $\mathcal{A}$ that reduces the Consensus problem to the frugal oracle with $k = 1$ [Anceaume et al. 2018]

1 **upon** propose(b) **do**
2      $validBlock \leftarrow \perp$
3      $validBlockSet \leftarrow \emptyset$ ▷ since $k = 1$ then it contains only one element
4   **while** $validBlock = \perp$ **do**
5          $validBlock \leftarrow$ getToken(b_0, b)
6      $validBlockSet \leftarrow$ consumeToken($validBlock$);   ▷ it can be different from validBlock
7   **trigger** decide($validBlockSet$)
8 **end upon**

---

at the beginning of the operation. In comparison with consumeToken($b_\ell^{tkn_h}$) we have that $b_\ell^{tkn_h}$ is the *new_value*, *register* is $K[h]$, and the implicit *old_value* is $\{\}$. That is, add(K, h, b) stores b in $K[h]$ if $|K[h]| < k = 1$, then if $K[h] = \{\}$. In any case the operation returns the content of $K[h]$ at the end of the operation itself. Algorithm 4.9 describes the pseudo-code that reduces CAS to consumeToken().

Theorem 4.24 ([Anceaume et al. 2018])   *If input values are in $\mathcal{B}'$, then there exists an implementation of CAS by CT in the case of $\Theta_{F,k=1}$.*

Algorithm 4.10 describes a simple implementation of Consensus by $\Theta_{F,k=1}$. When a correct process p_i invokes the propose(b) operation, it loops, invoking the getToken(b_0, b) operation as long as a valid block is returned. In this case the getToken() operation takes as input some block b_0 and the proposed block b. Afterwards, when the valid block has been obtained, p_i invokes the consumeToken($validBlock$) operation whose result is stored in the *tokenSet* variable. Notice that the first process that invokes this operation is able to successfully consume the token; that is, the valid block is in the oracle set corresponding to b_0, whose cardinality is $k = 1$, and such a set is returned each time the consumeToken() operation is invoked for a block related to b_0. Finally, the decision is triggered on such a set, which contains a single element.

Theorem 4.25 ([Anceaume et al. 2018]) $\Theta_{F,k=1}$ *oracle has Consensus number* ∞.

Theorem 4.26 ([Anceaume et al. 2018])   *There does not exist an algorithm* $\mathfrak{I}_{\Theta_{F,k=1}\to\mathcal{C}}$ *such that* $\Theta_{F,k=1}$ *oracle reduces to* $\mathcal{C}$.

Prodigal not stronger than an atomic register

In order to show that the prodigal oracle Θ_P has Consensus number 1, it suffices to find a wait-free implementation of the oracle by an object with Consensus number 1. To this end, the authors in Anceaume et al. [2018] propose a straightforward implementation of the prodigal oracle by Atomic Snapshot [Afek et al. 1990].

Let us first simplify the notation of the consume token operation. Let us consider a consume token invoked for a given block b_h, denoted in the following as consumeToken$_h(tkn_m)$, which simply writes a token from the set $\mathfrak{T} = \{tkn_1, tkn_2, \ldots, tkn_m, \ldots\}$ in the set $K[h]$. Without loss of generality, let us assume that: (i) tokens are uniquely identified, (ii) cardinality of $\mathfrak{T}$ is n finite but not known, and (iii) the set $K[h]$ is represented by a collection of n atomic registers $\mathfrak{K}[\mathfrak{h} = \{R_{h,1}, R_{h,2}, \ldots, R_{h,m}, \ldots R_{h,n}\}$, where $R_{h,m}$ is assigned to the tkn_m token—that is, $R_{h,m}$ can contain either $\perp$ or tkn_m.

It can be observed that the consumeToken$_h(tkn_m)$ in the case of k infinite always allows us to write the token tkn_m in $R_{h,m}$; that is, there always exists a register $R_{h,m}$ for the proposed token tkn_m. By the oracle definition, the consumeToken$_h(tkn_m)$ returns a read of the n registers that includes the last written token. Algorithm 4.11 shows a trivial implementation of commit token CT using Atomic Snapshot that offers both update(R_i, *value*), and scan($R_1, R_2, \ldots, R_n$) operations to update a particular register and perform an atomic read of input registers, respectively.

Theorem 4.27 ([Anceaume et al. 2018]) Θ_P *oracle has Consensus number* 1.

Algorithm 4.11: An implementation of CT by Atomic Snapshot in the case of Θ_P [Anceaume et al. 2018]

1 **function** consumeToken$_h(tkn)$
2      $R_{h,m} \leftarrow$ update($R_{h,m}, tkn_m$)
3   *returned_value* $\leftarrow$ scan($R_{h,1}, R_{h,2}, \ldots, R_{h,m}, \ldots R_{h,n}$)
4   **return** *returned_value*

Implementation of BT-ADT in message passing

We consider a message-passing system composed of an arbitrary large but finite set of n processes, $\Pi = \{p_1, \ldots, p_n\}$. The passage of time is measured by a fictional global clock (e.g., that spans the set of natural integers). Processes in the system do not have access to the fictional global time. Each process of the distributed system executes a single instance of a distributed protocol $\mathcal{P}$ composed of a set of algorithms—that is, each process is running an algorithm. Processes can exhibit a Byzantine behavior (i.e., they can arbitrarily deviate from the protocol $\mathcal{P}$ they are supposed to run). A process affected by a Byzantine behavior is said to be faulty; otherwise we say that this process is non-faulty or correct. We make no assumption on the number of failures that can occur during the system execution. Processes communicate by exchanging messages.

The BlockTree being now a shared object replicated at each process, we note by bt_i the local copy of the BlockTree maintained at process i. To maintain the replicated object, we consider histories made of events related to the read and append operations on the shared object—specifically the send and receive operations for process communications and the update operation for BlockTree updates. We also use subscript i to indicate that the operation occurred at process i: $\text{update}_i(b_g, b_i)$ indicates that i inserts its locally generated valid block b_i in bt_i with b_g as a predecessor. Updates are communicated through send and receive operations. An update related to a block b_i generated on a process p_i, sent through $\text{send}_i(b_g, b_i)$, and received through $\text{receive}_j(b_g, b_i)$, takes effect on the local replica bt_j of p_j with the operation $\text{update}_j(b_g, b_i)$.

We assume a generic implementation of the update operation: when process i locally updates its BlockTree bt_i with the valid block b_i (returned from the consumeToken() operation), we write $\text{update}_i(b, b_i')$. When a process j executes the $\text{receive}_j(b, b_i)$ operation, it locally updates its BlockTree bt_j by invoking the $\text{update}_j(b, b_i)$ operation.

In the remaining part of the work we consider implementations of BT-ADT in a Byzantine failure model where the set of events is restricted as follows.

Definition 4.28   The execution of the system that uses the BT-ADT $= (A, B, Z, \xi_0, \tau, \delta)$ in a Byzantine failure model defines the concurrent history $H = \langle \Sigma, E, \Lambda, \mapsto, \prec, \nearrow \rangle$ (see Definition 4.4) where we restrict E to a countable set of events that contains (i) all the BT-ADT read() operations invocation events by the *correct* processes, (ii) all BT-ADT read() operations response events at the *correct* processes, (iii) all append(b) operations invocation events such that b satisfies the predicate P, and (iv) send, receive, update operations events generated at correct processes.

Communication abstractions

We now define the properties that each history H generated by a BT-ADT (satisfying the eventual prefix property) has to satisfy, and then we prove their necessity.

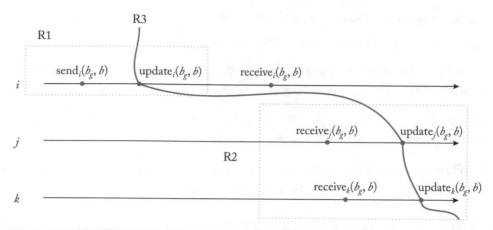

FIGURE 4.2: Example of a concurrent history that satisfies R1, R2, and R3, the Update Agreement properties [Anceaume et al. 2019b].

Definition 4.29 (Update Agreement)   A concurrent history $H = \langle \Sigma, E, \Lambda, \mapsto, \prec, \nearrow \rangle$ of the system that uses a BT-ADT satisfies the Update Agreement if it satisfies the following properties:

R1. $\forall \mathsf{update}_i(b_g, b_i) \in H, \exists \mathsf{send}_i(b_g, b_i) \in H$;

R2. $\forall \mathsf{update}_i(b_g, b_j) \in H, \exists \mathsf{receive}_i(b_g, b_j) \in H$ such that $\mathsf{receive}_i(b_g, b_j) \mapsto \mathsf{update}_i(b_g, b_j)$;

R3. $\forall \mathsf{update}_i(b_g, b_j) \in H, \exists \mathsf{receive}_k(b_g, b_j) \in H, \forall k$.

Figure 4.2 provides an example of a concurrent history that satisfies the three properties of the Update Agreement.

Theorem 4.30 ([Anceaume et al. 2019b])   *The Update Agreement property is necessary to construct concurrent histories $H = \langle \Sigma, E, \Lambda, \mapsto, \prec, \nearrow \rangle$ generated by a BT-ADT that satisfy the BT eventual consistency criterion.*

In Anceaume et al. [2019b], the authors conclude that there does not exist a concurrent history $H = \langle \Sigma, E, \Lambda, \mapsto, \prec, \nearrow \rangle$ of the system that uses a BT-ADT that satisfies the strong BT consistency criterion but not the Update Agreement property.

In the following we consider a communication primitive that is inspired by the liveness properties of the reliable broadcast [Cachin et al. 2011]. In Anceaume et al. [2019b], the authors prove that this abstraction is necessary to implement eventual BT consistency.

Definition 4.31 (Light reliable communication (LRC))   A concurrent history H satisfies the properties of the LRC abstraction if and only if:

- Validity: $\forall \mathsf{send}_i(b, b_i) \in H$, $\exists \mathsf{receive}_i(b, b_i) \in H$;

- Agreement: $\forall \mathsf{receive}_i(b, b_j) \in H$, $\forall k \exists \mathsf{receive}_k(b, b_i) \in H$.

In other words, if a correct process i sends a message m, then i eventually receives m, and if a message m is received by some correct process, them m is eventually received by every correct process.

Theorem 4.32 ([Anceaume et al. 2019b])   *The LRC abstraction is necessary for any BT-ADT implementation that generates concurrent histories that satisfy the BT eventual consistency criterion.*

It follows that the LRC abstraction is necessary in any BT-ADT implementation that generates concurrent histories that satisfy the BT strong consistency property.

4.4.5   HIERARCHY AND MAPPING WITH EXISTING SYSTEMS

This section presents the hierarchy of different BT-ADTs satisfying different consistency criteria when augmented with different oracle ADTs. Notations BT-ADT$_{SC}$ and BT-ADT$_{\diamond C}$ refer respectively to BT-ADTs generating concurrent histories that satisfy the strong consistency (SC) and the eventual consistency ($\diamond C$) criteria. When augmented with the oracles, four typologies are obtained, where for the *frugal* oracle an explicit value for k is given: $\mathfrak{R}(\text{BT-ADT}_{SC}, \Theta_{F,k})$, $\mathfrak{R}(\text{BT-ADT}_{SC}, \Theta_P)$, $\mathfrak{R}(\text{BT-ADT}_{\diamond C}, \Theta_P)$, and $\mathfrak{R}(\text{BT-ADT}_{\diamond C}, \Theta_{F,k})$.

In what follows the relationships among the different refinements are presented. Without loss of generality, let us consider only the set of histories $\hat{\mathcal{H}}^{\mathfrak{R}(\text{BT-ADT}, \Theta)}$ such that each history $\hat{H}^{\mathfrak{R}(\text{BT-ADT}, \Theta)} \in \hat{\mathcal{H}}^{\mathfrak{R}(\text{BT-ADT}, \Theta)}$ is purged from the unsuccessful append() response events (i.e., such that the returned value is $\bot$). Let $\hat{\mathcal{H}}^{\mathfrak{R}(\text{BT-ADT}, \Theta_F)}$ be the concurrent set of histories generated by a BT-ADT enriched with Θ_F-ADT and let $\hat{\mathcal{H}}^{\mathfrak{R}(\text{BT-ADT}, \Theta_P)}$ be the concurrent set of histories generated by a BT-ADT enriched with Θ_P-ADT.

Theorem 4.33 ([Anceaume et al. 2019b]) $\hat{\mathcal{H}}^{\mathfrak{R}(\text{BT-ADT}, \Theta_F)} \subseteq \hat{\mathcal{H}}^{\mathfrak{R}(\text{BT-ADT}, \Theta_P)}$.

Theorem 4.34 ([Anceaume et al. 2019b])   *If* $k_1 \leq k_2$, *then* $\hat{\mathcal{H}}^{\mathfrak{R}(\text{BT-ADT}, \Theta_{F,k_1})} \subseteq \hat{\mathcal{H}}^{\mathfrak{R}(\text{BT-ADT}, \Theta_{F,k_2})}$.

Theorem 4.35 ([Anceaume et al. 2019b]) $\hat{\mathcal{H}}^{\mathfrak{R}(\text{BT-ADT}_{SC}, \Theta)} \subseteq \hat{\mathcal{H}}^{\mathfrak{R}(\text{BT-ADT}_{\diamond C}, \Theta)}$.

Combining the previous theorems, we obtain the hierarchy depicted in Figure 4.3.

Thanks to Theorem 4.36 below, we can eliminate from the hierarchy in Figure 4.3 both $\mathfrak{R}(\text{BT-ADT}_{SC}, \Theta_P)$ and $\mathfrak{R}(\text{BT-ADT}_{SC}, \Theta_{F,k>1})$, since in both cases the Θ-ADT employed allows forks; thus such enriched ADTs cannot generate histories that satisfy the BT strong consistency criterion.

Theorem 4.36 ([Anceaume et al. 2019b])   *There does not exist an implementation of a BT-ADT that generates histories satisfying the BT strong consistency if forks occur.*

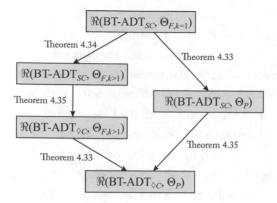

FIGURE 4.3: $\Re(\text{BT-ADT}, \Theta)$ hierarchy [Anceaume et al. 2019b].

TABLE 4.1: Mapping of existing systems. Each of these systems assumes at least a light reliable communication [Anceaume et al. 2019b]. (The abbreviation v.h.p. means "very high probability.")

References	Refinement
Bitcoin [Nakamoto 2008]	$\Re(\text{BT-ADT}_{\Diamond C}, \Theta_P)$
Ethereum [Wood 2017]	$\Re(\text{BT-ADT}_{\Diamond C}, \Theta_P)$
Algorand [Gilad et al. 2017]	$\Re(\text{BT-ADT}_{SC}, \Theta_{F,k=1})$ SC with v.h.p
ByzCoin [Kogias et al. 2016]	$\Re(\text{BT-ADT}_{SC}, \Theta_{F,k=1})$
PeerCensus [Decker et al. 2016]	$\Re(\text{BT-ADT}_{SC}, \Theta_{F,k=1})$
Red Belly [Crain et al. 2017]	$\Re(\text{BT-ADT}_{SC}, \Theta_{F,k=1})$
Hyperledger [Androulaki et al. 2018]	$\Re(\text{BT-ADT}_{SC}, \Theta_{F,k=1})$

Table 4.1 summarizes the mapping between different existing systems and these abstractions.

4.5   DISCUSSION

The work in Fernández Anta et al. [2019] extended the notion of a DLO (as presented in Section 4.3) by defining multi-distributed ledger objects (MDLOs), which are the result of aggregating multiple distributed ledger objects, used for the purposes of blockchain interconnection and interoperability. To demonstrate the use of MDLOs, the *Atomic Appends* problem was introduced, which emerges when the exchange of digital assets between multiple clients may involve appending records in more than one DLO. Specifically, the Atomic Appends problem requires that either all records will be appended on the involved DLOs or none. The solvability of the problem was examined under client crashes. Recently, in Cholvi et al. [2020], a formalization and implementation of a linearizable DLO tolerating Byzantine failures was introduced. Furthermore, the Atomic Appends problem was considered under Byzantine clients and servers. In all these works, the servers implementing the ledger are static. A challenging research direction is to deal with highly dynamic sets of possibly anonymous servers in order to implement distributed ledgers—to get closer, for example, to the Bitcoin-like ecosystem, and hence to permissionless blockchains. In this respect, relaxed versions of the strong prefix property required by the DLO formulation (which prevents forks) could be considered, following the approach of a weak prefix as considered in Section 4.4.

Another challenging future direction is to fully explore the properties of validated distributed ledgers and their relation with cryptocurrencies. The main challenge here is to devise appropriate *valid*() functions and implement them in an efficient manner.

The work in Anceaume et al. [2018] and [2019b], discussed in Section 4.4, presented an extended formal specification of blockchains and derived interesting conclusions on their implementability in both shared memory and message passing. Let us note that this work is intended to provide the groundwork for the construction of a sound hierarchy of blockchain abstractions and correct implementations. It did leave open several issues, however, such as the solvability of the Eventual prefix in message passing, the synchronization power of other oracle models, and fairness properties for oracles.

In the continuation of the work in Anceaume et al. [2018] and [2019b], the authors in Anceaume et al. [2020] have reinvestigated the definition of (deterministic) eventual prefix consistency presented in Anceaume et al. [2019b] to fit the context in which an infinite number of blocks are appended to the blockchain. In this work, they introduce the notion of bounded revocation, which informally says that the number of blocks that can be pruned from the current blockchain is bounded. Providing solutions that guarantee a known and bounded revocation turns out to be an important crux in the construction of blockchains. Specifically, in this work these authors show that known bounded revocation eventual finality is equivalent to Consensus, that unknown bounded revocation eventual finality—that is, eventual finality guaranteeing an unknown but bounded revocation—is equivalent to eventual immediate finality, and that eventual immediate finality is not weaker than eventual consensus, an abstraction that captures eventual agreement among all participants. They provide an algorithm that guarantees eventual finality in an asynchronous environment with an unbounded number of Byzantine processes, and show that it is impossible to build a blockchain that guarantees eventual finality when the rule to select a chain out of multiple ones (i.e., after a fork) is the longest chain rule. Finally, impossibilities and possibilities of unbounded revocation eventual finality are discussed. In particular, they propose an algorithm that solves unknown bounded revocation eventual finality in an eventually synchronous environment in the presence of less than a majority of Byzantine processes.

The work in Anceaume et al. [2017] was the first attempt to make the connection between distributed ledgers and the distributed shared objects (registers) theory. In Anceaume et al. [2017] the authors prove that Bitcoin and Ethereum verify only the specification of a distributed ledger register with regular semantics. Regular registers provide weak guarantees that are not easy for the application programmer to use for building correct applications on top of those registers. More significantly, they do not compose. Atomicity or linearizability is a state property that is composable [Herlihy 1990], and thus would allow the combination of multiple distributed ledgers, and has clear strong semantics. Atomicity has been adopted as the standard property to have in the development of parallel and distributed systems. Previous work [Anceaume et al. 2017] left open the question of proposing distributed ledger registers with atomic semantics. In Anceaume et al. [2019a], the authors have extended the work proposed in Anceaume et al. [2017] in several ways. First, in this work the authors propose a specification of a distributed ledger register that matches the Lamport hierarchy [Lamport 1986] from safe to atomic distributed ledger registers in a Byzantine prone environment. They propose implementations of the distributed ledger register that verify safe, regular, and atomic semantics, assuming the presence of Byzantine nodes. The model of communication is specific to distributed ledgers technology [Decker and Wattenhofer 2013]. Specifically, the underlying system provides a broadcast primitive satisfying the Δ-delivery property (if a node invokes the broadcast primitive with m as parameter then every correct node eventually delivers m once and within Δ time units). Finally, the authors propose an implementation of a distributed ledger register that satisfies the atomic specification and the k-consistency property defined in Anceaume et al. [2017] to mimic permissionless distributed blockchains. It should be noted that this work is complementary to the work discussed in Section 4.3 that focuses only on ledger objects that consist of a totally ordered sequence of blocks (or records). Unifying this framework with the one proposed in Fernández Anta et al. [2018] (i.e., Section 4.3) and extending it to multi-objects operations [Fernández Anta et al. 2019] is an interesting open direction. Moreover, connecting this framework with the runtime specification discussed in Section 4.4 in order to automatically design and verify distributed ledgers algorithms with various semantics is an interesting and important open research direction.

BIBLIOGRAPHY

Y. Afek, D. Dolev, H. Attiya, E. Gafni, M. Merritt, and N. Shavit. Atomic snapshots of shared memory. In *Proceedings of the 9th Annual ACM Symposium on Principles of Distributed Computing (PODC)*, pages 1–13. ACM, 1990. 121

M. Ahamad, G. Neiger, J. E Burns, P. Kohli, and P. W. Hutto. Causal memory: Definitions, implementation, and programming. *Distributed Computing* 9(1):37–49, 1995. 104

E. Anceaume, R. Ludinard, M. Potop-Butucaru, and F. Tronel. Bitcoin a distributed shared register. In *Proceedings of the International Symposium on Stabilization, Safety, and Security of Distributed Systems (SSS)*, 2017. 96, 127

E. Anceaume, A. Del Pozzo, R. Ludinard, M. Potop-Butucaru, and S. Tucci Piergiovanni. Blockchain abstract data type. *CoRR*, http://arxiv.org/abs/1802.09877, 2018. 118, 119, 120, 121, 126

E. Anceaume, M. Papatriantafilou, M. Potop-Butucaru, and P. Tsigas. Distributed ledger register: From safe to atomic. Preprint hal-02201472, https://hal.archives-ouvertes.fr/hal-02201472/file/main.pdf, July 2019a. 127

E. Anceaume, A. D. Pozzo, R. Ludinard, M. Potop-Butucaru, and S. Tucci Piergiovanni. Blockchain abstract data type. In C. Scheideler and P. Berenbrink, editors, *The 31st ACM Symposium on Parallelism in Algorithms and Architectures, SPAA 2019, Phoenix, Arizona, USA, June 22–24, 2019*, pages 349–358. ACM, 2019b. 96, 105, 109, 110, 111, 112, 113, 114, 115, 116, 117, 118, 123, 124, 125, 126

E. Anceaume, A. D. Pozzo, T. Rieutord, and S. Tucci-Piergiovanni. On finality in blockchains, 2020. https://hal-cea.archives-ouvertes.fr/cea-03080029. 126

E. Androulaki, A. Barger, V. Bortnikov, C. Cachin, K. Christidis, A. De Caro, D. Enyeart, C. Ferris, G. Laventman, Y. Manevich, S. Muralidharan, C. Murthy, B. Nguyen, M. Sethi, G. Singh, K. Smith, A. Sorniotti, C. Stathakopoulou, M. Vukolic, S. Weed Cocco, and J. Yellick. Hyperledger fabric: A distributed operating system for permissioned blockchains, 2018. https://arxiv.org/pdf/1801.10228v1.pdf. 125

H. Attiya and J. L. Welch. Sequential consistency versus linearizability. *ACM Trans. Comput. Syst.* 12(2):91–122, 1994. 103, 104

H. Attiya and J. Welch. *Distributed Computing: Fundamentals, Simulations and Advanced Topics*. John Wiley & Sons, 2004. 107

C. Cachin, R. Guerraoui, and L. E. T. Rodrigues. *Introduction to Reliable and Secure Distributed Programming (2nd ed.)*. Springer, 2011. 123

C. Cachin, K. Kursawe, F. Petzold, and V. Shoup. Secure and efficient asynchronous broadcast protocols. In *Advances in Cryptology—CRYPTO 2001, 21st Annual International Cryptology Conference, Santa Barbara, California, USA, August 19–23, 2001, Proceedings*, pages 524–541. Springer-Verlag, 2001. 108

S. Chaudhuri, R. Gawlick, and N. Lynch. Designing algorithms for distributed systems with partially synchronized clocks. In *Proceedings of the 12th Annual ACM Symposium on Principles of Distributed Computing*, PODC '93, New York, New York, USA, pages 121–132. ACM, 1993. 107

V. Cholvi, A. Fernández Anta, C. Georgiou, N. Nicolaou, and M. Raynal. Atomic appends in asynchronous Byzantine distributed ledgers. In *16th European Dependable Computing Conference, EDCC 2020, Munich, Germany, September 7–10, 2020*, pages 77–84. IEEE, 2020. 100, 105, 126

T. Crain, V. Gramoli, M. Larrea, and M. Raynal. (Leader/randomization/signature)-free Byzantine consensus for consortium blockchains, 2017. http://csrg.redbellyblockchain.io/doc/ConsensusRedBellyBlockchain.pdf. 108, 118, 119, 125

C. Decker, J. Seidel, and R. Wattenhofer. Bitcoin meets strong consistency. In *Proc. of the ICDCN International Conference*, Singapore, pages 1–10, 2016. https://doi.org/10.1145/2833312.2833321. 125

C. Decker and R. Wattenhofer. Information propagation in the Bitcoin network. In *13th IEEE International Conference on Peer-to-Peer Computing, IEEE P2P 2013, Trento, Italy, September 9–11, 2013, Proceedings*, pages 1–10. IEEE, 2013. 127

X. Défago, A. Schiper, and P. Urbán. Total order broadcast and multicast algorithms: Taxonomy and survey. *ACM Comput. Surv.* 36(4):372–421, 2004. 105

A. Fernández Anta, C. Georgiou, and N. Nicolaou. Atomic appends: Selling cars and coordinating armies with multiple distributed ledgers. In V. Danos, M. Herlihy, M. Potop-Butucaru, J. Prat, and S. T. Piergiovanni, editors, *International Conference on Blockchain Economics, Security and Protocols, Tokenomics 2019, Paris, France, May 6–7, 2019*, volume 71 of *OASIcs*, pages 5:1–5:16. Schloss Dagstuhl–Leibniz-Zentrum für Informatik, 2019. 125, 127

A. Fernández Anta, K. M. Konwar, C. Georgiou, and N. C. Nicolaou. Formalizing and implementing distributed ledger objects. *SIGACT News* 49(2):58–76, 2018. 96, 100, 105, 107, 127

J. A. Garay, A. Kiayias, and N. Leonardos. The Bitcoin backbone protocol: Analysis and applications. In *Advances in Cryptology—EUROCRYPT 2015—34th Annual International Conference on the Theory and Applications of Cryptographic Techniques*, Sofia, Bulgaria, 2015. 96

M. Gentz and J. Dude. Tunable data consistency levels in Microsoft Azure Cosmos DB, June 2017. https://docs.microsoft.com/en-us/azure/cosmos-db/consistency-levels. (Accessed 20 Oct. 2017.) 103

Y. Gilad, R. Hemo, S. Micali, G. Vlachos, and N. Zeldovich. Algorand: Scaling Byzantine agreements for cryptocurrencies. In *Proceedings of the 26th Symposium on Operating Systems Principles*, pages 51–68. ACM, 2017. 125

A. Girault, G. Gössler, R. Guerraoui, J. Hamza, and D.-A. Seredinschi. Monotonic prefix consistency in distributed systems. In *International Conference on Formal Techniques for Distributed Objects, Components, and Systems*, Berlin, Germany, pages 41–57, Springer, 2018. 96, 109, 110

M. Herlihy. Concurrency and availability as dual properties of replicated atomic data. *J. ACM* 37(2):257–278, 1990. 127

M. Herlihy. Wait-free synchronization. *ACM Transactions on Programming Languages and Systems (TOPLAS)* 13(1):124–149, 1991. 119

M. Herlihy. Blockchains and the future of distributed computing. In E. M. Schiller and A. A. Schwarzmann, editors, *Proceedings of the ACM Symposium on Principles of Distributed Computing, PODC 2017, Washington, DC, USA, July 25–27, 2017*, page 155. ACM, 2017. 95

M. P. Herlihy and J. M. Wing. Linearizability: a correctness condition for concurrent objects. *ACM Transactions on Programming Languages and Systems (TOPLAS)* 12(3):463–492, 1990. 103

E. K. Kogias, P. Jovanovic, N. Gailly, I. Khoffi, L. Gasser, and B. Ford. Enhancing Bitcoin security and performance with strong consistency via collective signing. In *25th USENIX Security Symposium*, 2016. 125

L. Lamport. How to make a multiprocessor computer that correctly executes multiprocess programs. *IEEE Transactions on Computers* C-28(9):690–691, 1979. 103, 104

L. Lamport. On inter-process communications, Part I: Basic formalism and Part II: Algorithms. *Distributed Computing* 1(2):77–101, 1986. 127

L. Lamport, R. Shostak, and Marshall Pease. The Byzantine generals problem. *ACM Transactions on Programming Languages and Systems* 4:382–401, 1982. 118

M. Mavronicolas and D. Roth. Linearizable read/write objects. *Theor. Comput. Sci.* 220(1):267–319, 1999. 107

S. Nakamoto. Bitcoin: A peer-to-peer electronic cash system, 2008. https://bitcoin.org/en/bitcoin-paper. (Accessed 3 Apr. 2018.) 108, 125

R. Pass and E. Shi. Fruitchains: A fair blockchain. In *Proceedings of the ACM Symposium on Principles of Distributed Computing, PODC 2017*, New York, New York, USA, pages 315–324. ACM, 2017. 96

M. Perrin. *Distributed Systems, Concurrency and Consistency*. ISTE Press, Elsevier, 2017. 97, 99

M. Perrin, Achour Mostefaoui, and Claude Jard. Causal consistency: Beyond memory. In *21st ACM SIGPLAN Symposium on Principles and Practice of Parallel Programming*, Barcelona, Spain, pages 1–12, 2016. https://doi.org/10.1145/3016078.2851170. 109

M. Raynal. *Concurrent Programming: Algorithms, Principles, and Foundations*. Springer, 2013. 97, 105

J. Wang, E. Talmage, H. Lee, and J. L. Welch. Improved time bounds for linearizable implementations of abstract data types. In *2014 IEEE 28th International Parallel and Distributed Processing Symposium*, Phoenix, Arizona, pages 691–701. May 2014. 107

G. Wood. Ethereum: A secure decentralised generalised transaction ledger, 2017. http://gavwood.com/Paper.pdf. 125

AUTHORS' BIOGRAPHIES

E. Anceaume is a senior researcher scientist of CNRS at the IRISA laboratory (CNRS is the French national research agency). She holds a PhD in Computer Science from the University Paris-Orsay (Paris-XI) for her work on dependable systems. She spent one year as a postdoc researcher at Cornell University (USA). She is interested in dependability and security issues in large scale and dynamic distributed systems. For the last few years, she has been working on data stream algorithms, reputation mechanisms, dependability issues in peer-to-peer systems, and on distributed ledgers (blockchains).

Antonio Fernández Anta is Research Professor at IMDEA Networks. Previously he was on the faculty of the Universidad Rey Juan Carlos (URJC), and the Universidad Politécnica de Madrid

(UPM), where he received a research performance award. He was a postdoc at MIT (1995–1997), and spent sabbatical years at Bell Labs and MIT Media Lab. He has been awarded the Premio Nacional de Informática "Aritmel" in 2019 and has been a Mercator Fellow of the SFB MAKI in Germany since 2018. He received his MSc and PhD from the University of Louisiana. He is a Senior Member of ACM and IEEE.

Chryssis Georgiou is an Associate Professor in the Department of Computer Science at the University of Cyprus. He holds a PhD (2003) and MSc (2002) in Computer Science and Engineering from the University of Connecticut. His research interests span the theory and practice of fault-tolerant distributed computing with a focus on algorithms and complexity. Recent research topics include the specification and implementation of distributed ledgers, the design and implementation of fault-tolerant and strongly consistent distributed storage systems, the design and analysis of self-stabilizing distributed systems, and the application of crowdsourcing to fight against the spread of COVID-19.

Nicolas Nicolaou is a co-founder and a senior scientist and algorithms engineer at Algolysis Ltd. He held various academic positions—as visiting faculty until 2014, as an IEF Marie Curie Fellow at IMDEA Networks Institute (2014–2016), as a short-term scholar at MIT (2017), and as a postdoc researcher at the KIOS Research Center of Excellence (2017–2019)—before departing to an industrial position in 2019. He holds a PhD (2011) and a MSc (2006) from the University of Connecticut and a BSc (2003) from the University of Cyprus. His main research interests lie in the areas of distributed systems, design and analysis of fault-tolerant distributed algorithms, distributed ledgers (blockchains), security for embedded devices and critical infrastructures, and sensor networks.

Maria Potop-Butucaru is a full professor at Sorbonne University and a researcher in LIP6 laboratory (Laboratoire de recherche en informatique de Paris 6). Her research focuses on distributed systems resilient to multi-faults and attacks (crash, Byzantine, transient, etc.). She is interested in self*(self-organizing, self-healing, and self-stabilizing) and secure static and dynamic distributed systems (e.g., blockchains, peer-to-peer networks, sensors, and robot networks). She focuses in particular on the conception and proof of dependable distributed algorithms for fundamental distributed computing problems: communication primitives (e.g., broadcast, converge-cast, etc.), self* overlays (various spanning trees, P2P overlays, etc.), coherence and resource allocation problems (storage, mutual exclusion, etc.), consensus or leader election.

CHAPTER 5

Adversarial Cross-Chain Commerce

Maurice Herlihy, *Brown University, Providence, RI, USA*
Barbara Liskov, *MIT, Cambridge, MA, USA*
Liuba Shrira, *Brandeis University, Waltham, MA, USA*

5.1   INTRODUCTION

Imagine that Bob, a theater owner, decides to sell two coveted tickets to a hit play for 100 coins. Alice, a broker, knows that Carol would be willing to pay 101 coins for those tickets, so Alice plans to resell Bob's tickets to Carol at a small mark-up. How can one devise a distributed protocol, to be executed by Alice, Bob, and Carol, that transfers the tickets from Bob to Carol, and the coins from Carol to Bob, minus Alice's commission? If all goes as planned, all transfers take place, and if anything goes wrong (someone crashes or tries to cheat), no honest party should end up worse off. For example, Alice should not end up holding tickets she can't sell or coins that she must refund.

This chapter explores how to think about multi-party deals that span multiple blockchains. There are several important questions to consider.

What does cross-chain commerce look like?
This chapter introduces *cross-chain deals*, a distributed commerce model that encompasses and generalizes current practice.

What kinds of atomicity and integrity guarantees should cross-chain commerce support?
Although cross-chain deals resemble classical distributed atomic transactions, they are not the same. For example, the classical "all-or-nothing" property of atomic transactions is impossible to guarantee if participants can behave in arbitrary, even seemingly irrational ways. This chapter describes safety and liveness properties to replace the classical notions of transactional atomicity.

How can these properties be implemented?
This chapter describes two approaches to implementing cross-chain deals: a fully decentralized *timelock* protocol that assumes a synchronous communication model, and a more centralized *certified blockchain* (CBC) protocol that does not.

Specifying correctness for systems in which parties can deviate arbitrarily from the common protocol requires care. One cannot even assume parties will be rational, because they may have unknown objective functions (such as a foreign power willing to pay to disrupt an economy). Furthermore, some familiar classical properties such as the "all-or-nothing" property of atomic transactions cannot be enforced in an adversarial setting. In much of the literature on cross-chain swaps and transactions, correctness is treated informally, or in an ambiguous way. Without a realizable (and realistic) notion of correctness, it is impossible to reason about the correctness of blockchain protocols, smart contract code, or any other subsystem supporting adversarial commerce.

5.2   SYSTEM MODEL

5.2.1   TERMINOLOGY

We use *blockchain* in a generic sense to mean a publicly readable, tamper-proof distributed ledger (or database) that tracks ownership of *assets* among various *parties*. Our treatment is largely independent of the particular algorithms used by participating blockchains.

An asset may be *fungible*, like cryptocurrency, or *non-fungible*, like the title to a building. A party can be a person, an organization, or even a contract (see below). There are multiple independent blockchains, each managing a different asset. We are concerned here with blockchains that track asset ownership, and with commerce where parties agree to exchange ownership of assets, perhaps in complicated ways. We assume all asset transfers in a deal are explicitly represented on the blockchain. For example, Alice does not send cryptocurrency to Bob in exchange for paper tickets.

Asset ownership is managed by simple programs called *smart contracts* (*contracts*, for short). (The name is historical; smart contracts are not contracts in any meaningful sense.) Like an object in an object-oriented programming language, a smart contract encapsulates long-lived state, and exports a set of *functions* through which parties can access that state. Both contract state and contract code reside on the blockchain, where they can be read by any party, ensuring that a party calling a contract knows what code will be executed. Contract code must be deterministic because many blockchain algorithms require contracts to be re-executed multiple times by mutually suspicious parties.

We will mostly use contracts for *escrow*: an asset owner temporarily transfers ownership of an asset to a contract. If certain conditions are met, the contract transfers that asset to a *counterparty*, and otherwise it refunds that asset to the original owner.

We say that a party *publishes* an entry on a blockchain when it updates public data on that blockchain (through a contract). A party *monitors* a blockchain if it is notified when another party publishes an entry of interest.

Contracts have one essential limitation that lies at the heart of why cross-chain commerce is difficult. A contract can read data and call other contracts on the blockchain where it resides, but it cannot directly observe data or call contracts on other blockchains. A contract on blockchain *A* can learn of a change to data on blockchain *B* only if some party explicitly informs *A* of *B*'s change. Because parties do not trust one another, it may be necessary to provide blockchain *A* with some kind of "proof" that the information about *B*'s state is correct.

5.2.2 FAULT MODEL

Parties to a deal are expected to follow a common *protocol*. Instead of distinguishing between faulty and non-faulty parties, as in classical models, we distinguish only between *compliant* parties who follow the protocol, and *deviating* parties who do not. Many kinds of fault-tolerant distributed protocols require that some fraction of the parties be compliant. For example, proof-of-work consensus [Nakamoto 2009] requires a compliant majority, while most Byzantine fault-tolerant (BFT) consensus protocols require more than two-thirds of the participants to be compliant. For cross-chain deals, however, it seems prudent to make no assumptions about the number of deviating parties.

In summary, contract code is passive, public, deterministic, and trusted, while parties are active, autonomous, and potentially dishonest.

5.2.3 TIMING

There are three timing models commonly used in the blockchain literature.

Synchronous model

There is a known upper bound on the propagation time for one party's change to the blockchain state to be noticed by the other parties. Proof-of-work protocols such as Bitcoin [Nakamoto 2009] and Ethereum [Ethereum 2021] use this model.

Semi-synchronous model

There is initially no bound on message propagation time, but the system eventually reaches a *global stabilization time* (GST) after which the system becomes synchronous. (In practice, the synchronous periods need only last "long enough" to stabilize the protocol.) Byzantine fault-tolerant consensus protocols such as Algorand [Gilad et al. 2017], Libra [Libra Association 2019], and Hot-Stuff [Abraham et al. 2018] operate in this model.

Asynchronous model

There is no bound on message propagation time. Deterministic consensus is impossible in such a model, but blockchains based on randomized consensus protocols, such as HoneyBadger [Miller et al. 2016], are possible.

Here, we focus on protocols that assume either the synchronous or semi-synchronous models, because these are the most common.

5.2.4 CRYPTOGRAPHIC MODEL

We make standard cryptographic assumptions. Each party has public and private keys, and public keys are known to all. Messages are signed and cannot be forged, and they include single-use labels ("nonces") so they cannot be replayed.

5.3 CROSS-CHAIN DEALS

Here we describe cross-chain deals, what it means to execute them, and what it means for them to be correct.

5.3.1 SPECIFYING THE DEAL

Each payoff (set of final transfers) for a deal can be expressed as a matrix (or table), where each row and column is labeled with a party, and the entry at row i and column j shows the assets to be transferred from party i to party j. A party's column states what it expects to acquire from the deal (its *incoming* assets), and its row states what it expects to relinquish (its *outgoing* assets). A party enters a deal if the proposed transfers leave it better off, and it agrees to commit (complete) the deal if it deems the actual payoff to be acceptable.

In our running example, the payoff is given by the 3×3 matrix in Table 5.1. Carol expects to transfer 101 coins to Alice in return for tickets transferred from Alice. Similarly, Bob expects to transfer tickets to Alice in return for 100 coins from Alice. Although the table refers only to "tickets,"

TABLE 5.1: Alice, Bob, and Carol's deal. Rows represent outgoing transfers, and columns incoming transfers.

	Alice	Bob	Carol
Alice		100 coins	tickets
Bob	tickets		
Carol	101 coins		

the specific (non-fungible) tickets to be provided would be part of the deal specification, while the specific (fungible) coins would likely be omitted.

A deal where Alice auctions an asset to Bob and Carol would require two matrices, one for each successful outcome: (1) if Bob outbids Carol, Alice transfers her asset to Bob, Bob transfers his bid to Alice, and Carol transfers her bid back to herself, and (2) if Carol outbids Bob, the transfers are symmetric. (A realistic on-chain auction would also include fees and deposits to penalize malicious behavior by bidders.)

5.3.2   STATE MACHINE MODEL

More formally, a cross-chain deal is a simple state machine that tracks ownership of assets, and whose transitions represent escrows, transfers, commits, and aborts.

Let $\mathcal{P}$ be a domain of *parties*, and $\mathcal{A}$ a domain of *assets*. (A party may be a person or a contract, and assets are digital tokens representing items of value.) An asset has exactly one *owner* at a time: $Owns(P, a)$ is *true* if P and only P owns a.

An active deal tentatively transfers asset ownership from one party to another. We say a tentative transfer *commits* if it becomes permanent, and it *aborts* if it is discarded. A deal *commits* if all its tentative transfers commit, and it *aborts* if all its tentative transfers abort.

While a deal is in progress, its state encompasses two maps, $C : \mathcal{A} \rightarrow \mathcal{P}$ and $A : \mathcal{A} \rightarrow \mathcal{P}$, both initially empty. $C(a)$ indicates the eventual owner of asset a if the deal commits at a's blockchain, and $A(a)$ the owner if it aborts at that blockchain. We use $Owns_C(P, a)$ to indicate that P will own a if the deal commits, and $Owns_A(P, a)$ to indicate that P will own a if the deal aborts.

Escrow plays the role of classical concurrency control, ensuring that a single asset cannot be transferred to different parties at the same time. Here is what happens when P places a in escrow during deal D:

$$\text{Pre:} \quad Owns(P, a)$$

$$\text{Post:} \quad Owns(D, a) \text{ and } Owns_C(P, a) \text{ and } Owns_A(P, a)$$

The precondition states that P can escrow a only if P owns a. If that precondition is satisfied, the postcondition states that ownership of a is transferred from P to D (via the escrow contract), but P remains the owner of a in both C and A, since no tentative transfer has happened yet, so P would regain ownership of a if D were to terminate either way. For example, when Bob escrows his tickets, they become the property of the contract, but should the deal terminate right then, the tickets would revert to Bob.

Next we define what happens when party P tentatively transfers an asset (or assets) a to party Q as part of deal D.

$$\text{Pre:} \quad Owns(D, a) \text{ and } Owns_C(P, a)$$

$$\text{Post:} \quad Owns_C(Q, a) \text{ and } Owns_A(P, a)$$

The precondition requires a to be held in escrow by D, with P the indicated owner should D commit. If the precondition is satisfied, the postcondition states that Q will become the owner of the transferred a should D commit (at this point). For example, when Carol transfers 101 coins to Alice, Alice becomes the owner of those coins in C. Alice can then transfer 100 of those coins to Bob, retaining one for herself, all in C.

Assets remain in escrow until the deal terminates. If the deal terminates by committing, the owners of assets in C become the actual owners (displacing D). If it terminates by aborting, the owners of assets in A become the actual owners (again displacing D).

5.3.3   PHASES
A deal is executed in the following phases.

Clearing phase.   A market-clearing service discovers and broadcasts the participants, the proposed transfers, and possibly other deal-specific information. The market-clearing service may be centralized, but *it is not a trusted party*, because each party later decides for itself whether to participate. The precise structure of the service is beyond the scope of this chapter.

Escrow phase.   Parties escrow their outgoing assets. For example, Bob escrows his tickets and Carol her coins.

Transfer phase.   The parties perform the sequence of tentative ownership transfers according to the deal. For example, Bob tentatively transfers the tickets to Alice, who subsequently transfers them to Carol.

Validation phase.   Once the tentative transfers are complete, each party checks that the deal is the same as proposed by the (untrusted) market-clearing service, that its incoming assets are properly escrowed (so they cannot be double-spent), and that the payoff defined by the incoming and outgoing assets is acceptable. For example, Carol checks that the tickets to be transferred are escrowed, that the seats are (at least as good as) the ones agreed upon, and that she is not about to somehow overpay.

In the classical two-phase commit protocol [Bernstein et al. 1986], validation usually requires no semantic checks; instead, a party agrees to prepare if appropriate locks are held and persistence is guaranteed. Under adversarial commerce, however, an application-specific validation phase is needed for each party to decide whether the proposed payoff is acceptable. For example, only Carol can decide whether the tickets she is about to purchase are ones she wants.

Commit phase.   The parties vote on whether to make the tentative transfers permanent. If all parties vote to commit, the escrowed assets are transferred to their new owners; otherwise they are refunded to their original owners.

Cross-chain deals rely on two critical, intertwined mechanisms. First, the escrow mechanism prevents double spending by making the escrow contract itself the asset owner. Care must be taken that assets belonging to compliant parties do not remain escrowed forever in the presence of malicious behavior by counterparties. Second, the commit protocol must be resilient in the presence of malicious misbehavior. A deviating party may be able to steal assets if it can convince some parties that the deal completed successfully, and others that it did not. If a deviating party can prevent (or delay) a decision by the commit protocol, then it can keep assets locked up forever (or a long time).

The principal challenge in implementing cross-chain deal protocols is the design of the integrated escrow management and commit protocol. Just as with classical transaction mechanisms, there are many possible choices and trade-offs. In the remainder of this chapter, we describe two cross-chain deal protocols, implemented via contracts: one for the synchronous timing model, and one for the semi-synchronous model, each making different trade-offs concerning decentralization and fault-tolerance.

5.4   CORRECTNESS

Parties to a deal agree on a *protocol* to complete the deal's transfers. In an environment where we cannot force the parties to follow a protocol, it is impossible to guarantee that all transfers take place as promised by the deal specification. Which kind of partial transfers should be deemed acceptable?

The most fundamental safety property is (informally) that compliant parties should end up "no worse off," even when other parties deviate arbitrarily from the protocol. A party's *payoff* for a protocol execution is the sets of incoming and outgoing assets actually transferred. Some payoffs are considered *acceptable*, the rest not. Some acceptable payoffs are preferable to others, but any acceptable payoff leaves that party "no worse off."

Every party considers the following payoffs acceptable: ALL, where all agreed transfers take place, and NOTHING, where no transfers take place. In addition, we allow a party to consider other payoffs acceptable. For example, a party that expects three incoming transfers and three outgoing transfers may be willing to accept a payoff where it receives only two incoming transfers in return for only two outgoing transfers. Of course, any such choice is application-dependent.

We also assume that if a payoff is acceptable to a party then so is any payoff where that party acquires even more incoming assets (it gets something for nothing), or it relinquishes strictly fewer outgoing assets (discount pricing). For example, a payoff where a party transfers no outgoing assets but receives some incoming assets is an acceptable modification (to that party) to the baseline NOTHING payoff. Such outcomes, while unlikely in practice, cannot be excluded.

Cross-chain task protocols typically rely on some form of *escrow* to ensure the good faith of participating parties. Yet under a correct protocol, conforming parties' assets cannot be locked up forever. Nor, ideally, are any parts of the transaction to be excluded. Correctness thus encompasses the following safety and liveness properties:

Property 5.1 (Safety)   For every protocol execution, every compliant party ends up with an acceptable payoff. This notion of safety replaces the classical *all-or-nothing* property of atomic transactions, which, as noted, cannot be implemented in the presence of deviating parties.

Property 5.2 (Weak liveness)   No asset belonging to a compliant party is escrowed forever.

Property 5.3 (Strong liveness)   If all parties are compliant and willing to accept their proposed payoffs, then all transfers happen (all parties' payoffs are ALL). It is a well-known result [Fischer et al. 1985] that strong liveness is possible only in periods when the communication network is synchronous, ensuring a fixed upper bound on message delivery time.

5.5   EXECUTION MODEL

Before we can discuss and compare possible deal implementations, it is beneficial to review Section 5.3.3's breakdown of deal execution into the following phases.

Clearing phase.   As noted, we assume there is some kind of market-clearing service that allows the participants to enroll in the proposed deal.

Escrow phase.   Parties escrow their outgoing assets. For example, Bob escrows his tickets and Carol her coins.

Transfer phase.   The parties perform the sequence of tentative ownership transfers according to the deal. For example, Bob tentatively transfers the tickets to Alice, who subsequently transfers them to Carol.

Validation phase.   Once the tentative transfers are complete, each party checks that the deal is the same as proposed by the (untrusted) market-clearing service, that its incoming assets are properly escrowed (so they cannot be double-spent), and that the payoff defined by the incoming and outgoing assets is acceptable. For example, Carol checks that the tickets to be transferred are escrowed, that the seats are (at least as good as) the ones agreed upon, and that she is not about to overpay.

Commit phase.   The parties vote on whether to make the tentative transfers permanent. If all parties vote to commit, the escrowed assets are transferred to their new owners; otherwise they are refunded to their original owners.

5.6   TIMELOCK PROTOCOL

We now describe a *timelock* commit protocol where escrowed assets are released when and if all parties vote to commit. Parties do not explicitly vote to abort. Instead, timeouts are used to ensure that escrowed assets are not locked up forever if some party crashes or walks away from the deal. This protocol assumes a *synchronous* network model where blockchain propagation time is known and bounded.

In our example, Bob places his tickets into escrow, then transfers them to Alice, who transfers them to Carol. All parties examine their incoming assets, and if the resulting payoff is acceptable, the parties vote to commit at the escrow contract on each asset's blockchain. For example, if Alice, Bob, and Carol all register commit votes on the ticket blockchain, the escrow contract releases the tickets to Carol. All votes are subject to timeouts: if any commit vote fails to appear before the contract's timeout expires, the tickets revert to Bob. (Symmetric conditions apply to Carol's coins.)

Because of the adversarial nature of a deal, each party is motivated to publish its vote on the blockchains controlling its incoming assets (it is eager to be paid), but not on the blockchains controlling its outgoing assets (it is not so eager to pay). To align the protocol with incentives, one party's commit vote may be *forwarded* from one escrow contract to another by a motivated party.

For example, Bob is motivated to publish his commit vote only on the coin blockchain. However, once published, Bob's vote becomes visible to Carol, who is motivated to forward that vote to the ticket blockchain. Carol's position is symmetric: she is motivated to publish her vote only on the ticket blockchain, but Bob is motivated to forward it to the coin blockchain. Alice is motivated to send her vote to both blockchains. (Nevertheless, no harm occurs if a party sends its commit vote directly to any contract.)

A tricky part of this protocol is how to choose timeouts. A protocol implementation that simply assigns each party a timeout for each asset does not satisfy our notions of correctness, as shown by the following example.

Suppose that the ticket and coin escrows assign Alice timeouts A_t and A_c respectively, and that Bob and Carol's commit votes have already been published on both blockchains. In one scenario, Alice waits until just before A_c to register her vote on the coin blockchain, unlocking Carol's payment to Bob. It may take time Δ for Carol to observe Alice's vote and forward it to the ticket blockchain, implying that $A_t \geq A_c + \Delta$. In another scenario, Alice waits until just before A_t to register her vote on the ticket blockchain, unlocking Bob's tickets for Carol. It may take time Δ for Bob to observe Alice's vote and forward it to the coin blockchain, implying that $A_c \geq A_t + \Delta$, a contradiction.

To resolve this dilemma, each escrow contract's timeout for a party's commit vote depends on the length of the path along which that vote was forwarded. For example, if Alice votes directly, her vote will be accepted only if it is received within Δ of the commit protocol's starting time. This vote must be signed by Alice. If Alice forwards a vote from Bob, that vote will be accepted only if it is received within $2 \cdot \Delta$ of the starting time, where the extra Δ reflects the worst-case extra time needed to forward the vote. This vote must be signed first by Bob, then Alice. Finally, if Alice forwards a vote that Bob forwarded from Carol, that vote will be accepted only if it is received within $3 \cdot \Delta$, and so on. This vote must be signed first by Carol, then Bob, then Alice. We refer to this chain of signatures as the vote's *path signature*.

In general, a vote from party X received with path signature p must arrive within time $|p| \cdot \Delta$ of the pre-established commit protocol starting time, where $|p|$ is the number of distinct signatures for that vote.

5.6.1 RUNNING THE PROTOCOL

Here is how to execute the phases of a timelock protocol.

Clearing phase. The market-clearing service broadcasts the following to all parties in the deal: the deal identifier D, the list of parties *plist*, a commit phase starting time t_0 used to compute timeouts, and the timeout delay Δ. Most blockchains measure time imprecisely, usually by multiplying the current block height by the average block rate. The choice of t_0 should be far enough in the future to take into account the time needed to perform the deal's tentative transfers, and Δ should be large enough to render irrelevant any imprecision in blockchain timekeeping. Because t_0 and Δ are used only to compute timeouts, their values do not affect normal execution times, where all votes are received in a timely way. If deals take minutes (or hours), then Δ could be measured in hours (or days).

Escrow phase. Each party places its outgoing assets in escrow through an escrow contract

$$escrow(D, Dinfo, a)$$

on that asset's blockchain. Here D is the deal identifier and *Dinfo* is the rest of the information about the deal (*plist*, t_0, and Δ); the escrow requests take effect only if the party is the owner of a and a member of the *plist*.

Transfer phase. Party P transfers an asset (or assets) a tentatively owned by P to party Q by sending

$$transfer(D, a, Q)$$

to the escrow contract on the asset's blockchain. The party must be the tentative owner of a and Q must be in the *plist*.

Validation phase. Each party examines its escrowed incoming assets to see if they represent an acceptable payoff and the deal information provided by the market-clearing service is correct. If so, the party votes to commit.

Commit phase. Each compliant party sends a commit vote to the escrow contract for each incoming asset. (A compliant party is free to altruistically send commit votes to other escrow contracts as well.) A party uses

$$commit(D, v, p)$$

to vote directly and to forward votes to the deal's escrow contracts, where v is the voter and p is the path signature for v's vote. For example, if Alice is forwarding Bob's vote, then v is Bob, and p contains first Bob's signature, and then Alice's signature. (Throughout, we assume that deal identifiers are unique to guard against replay attacks.)

A contract accepts a commit vote only if it arrives in time and is well-formed: all parties in the path signature are unique and in the *plist*, and their signatures are valid and attest to a vote from v. If the commit is accepted, that contract has now accepted a vote from the party.

A contract releases the escrowed asset to the new owner(s) when it accepts a commit vote from every party. If the contract has not accepted a vote from every party by time $t_0 + N \cdot \Delta$, where N is the number of parties, it will never accept the missing votes, so the contract times out and refunds its escrowed assets to the original owners.

5.6.2 WELL-FORMED DEALS AND DECENTRALIZATION

For ease of exposition, we can think of a deal as a *directed graph* (digraph), where each vertex represents a party, and each arc represents a transfer; the digraph for our deal is shown in Figure 5.1.

If the deal digraph is not strongly connected, it can be shown that the deal is not *well-formed*, in the sense that it must include one or more "free riders" that collectively take assets but do not

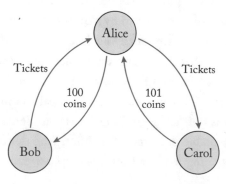

FIGURE 5.1: Alice, Bob, and Carol's deal expressed as a digraph.

return any [Herlihy 2018]. The remaining parties have no incentive to conform to any protocol executing such a deal, because they could improve their payoffs by excluding the free riders.[1]

A compliant party first sends votes to the escrow contracts on its incoming assets' blockchains. Then it monitors its outgoing assets' blockchains and forwards other parties' votes to its incoming assets' blockchains. No party needs to interact with any other blockchains. For example, if Carol owns only altcoins, then as part of the deal, she can go to David to exchange her altcoins for coins, and the deal can commit without parties such as Bob needing to interact with the altcoin blockchain (or even know about it). This protocol is *decentralized* in the sense that there is no single blockchain that must be accessed by all compliant parties.

This voting protocol reflects the (incentive-compatible) *minimum* a compliant party must do. Nothing prevents compliant parties from sending their commit votes directly to arbitrary blockchains, although typically parties will want to restrict themselves to blockchains they already use. For example, Bob might send his vote directly to the ticket blockchain, perhaps hoping to speed up the commit process. If he does so, he passes up a (very unlikely) opportunity to cheat Carol if she (non-compliantly) fails to claim her tickets in time.

In the remainder of this section, we assume that deals are well-formed, and the corresponding digraphs are strongly connected, although the timelock protocol can handle ill-formed deals if needed.

5.6.3   WHAT COULD POSSIBLY GO WRONG?

What can happen if parties deviate from the protocol? For example, suppose Bob wants to trade b-coins for c-coins, and Carol wants the reverse. Alice brokers their deal, taking 101 b-coins from Bob and 101 c-coins from Carol, then forwarding 100 of the opposite coins to each counterparty, keeping a 1-coin commission from each side. Suppose, moreover, that Alice happens to hold both kinds of coins. To save time, Alice performs multiple escrows in parallel. While Carol puts 101 of her c-coins in escrow for Alice, Alice puts 100 of her own c-coins in escrow for Bob, and similarly for Bob's b-coins. Now suppose Alice is infected by malware that causes her to behave irrationally. All parties vote to commit, but Alice irrationally neglects to forward her own and Bob's commit votes to the contract holding Bob's b-coins intended for Alice. Bob's timelock will eventually expire, and he will take back his coins, so for him the deal aborted. Carol, however, will transfer her 101 c-coins to Alice and receive her 100 b-coins, so for her the deal committed normally.

Although this outcome is not "all-or-nothing," it is considered acceptable because all compliant parties end up with acceptable outcomes. But Alice, who deviated from the protocol, foots the bill, paying Carol without being paid by Bob. We emphasize the unenforceability of classical

---

[1] Perhaps the free riders are sending some kind of hidden off-chain payments to the other parties, but support for hidden payments is beyond this chapter's scope.

correctness properties because it may seem counter-intuitive. But if we want to verify that contracts are correct, we must take care to use a realizable notion of correctness.

5.6.4   CORRECTNESS

Theorem 5.4   *The timelock protocol satisfies safety.*

Proof. By construction, transferring a compliant party X's escrowed incoming and outgoing assets is an acceptable payoff for X. Suppose by way of contradiction that X's outgoing asset a is released from escrow and transferred (with commit votes from every party), but the escrow for X's incoming asset b times out and is refunded because of a missing vote from party Z. Suppose Z's commit vote at a's contract arrived with path signature p. The signatures in p cannot include X's, because X is compliant and would have already forwarded Z's vote to b. Z's vote must have arrived at a before time $t_0 + |p| \cdot \Delta$. Since X is compliant, it forwards that vote to b's contract before time $t_0 + (|p| + 1) \cdot \Delta$, where that vote is accepted, a contradiction.   □

Theorem 5.5   *The timelock protocol satisfies weak liveness: no compliant party's outgoing assets are locked up forever.*

Proof. Every escrow created by a compliant party has a finite timeout.   □

Theorem 5.6   *The timelock protocol satisfies strong liveness.*

Proof. If all parties are compliant, they all send commit votes to the escrow contracts for their incoming assets. Each time a new commit vote appears on an outgoing asset's contract, the party forwards it to its incoming assets' contracts. Since the deal is well-formed, the deal digraph is strongly connected, and all commit votes are forwarded to all contracts in time.   □

Suppose that Bob acquires Alice and Carol's votes on time, and forwards them to claim the coins, but Alice and Carol are driven off-line before they can forward Bob's vote to the ticket blockchain, so Bob ends up with both the coins and the tickets. Technically, Alice and Carol have deviated from the protocol by not claiming their assets in time. As a countermeasure, Δ should be chosen large enough to make sustained denial-of-service attacks prohibitively expensive. For similar reasons, the Lightning payment network [Poon and Dryja 2016] employs *watchtowers* [Chester 2018] to monitor escrow contracts to act on the behalf of off-line parties.

5.7   CBC PROTOCOL

Now we describe a commit protocol that assumes only a semi-synchronous communication model [Dwork et al. 1988]. Since we cannot use timed escrow, we allow parties to vote to abort if validation fails, or if too much time has passed.

Unlike in the classical two-phase commit protocol [Bernstein et al. 1986], there is no coordinator; instead we use a special blockchain, the *certified blockchain*, or *CBC*, as a kind of shared log. The CBC might be a stand-alone blockchain or one of those already being used in the deal.

Instead of voting on individual assets, each party votes on the CBC whether to commit or abort the entire deal. The CBC records and orders these votes. A party can extract a *proof* from the CBC that particular votes were recorded in a particular order. A party claiming an asset (or a refund) presents a proof of commit (or abort) to the contract managing that asset. The contract checks the proof's validity and carries out the requested transfers if the proof is valid. A *proof of commit* proves that every party voted to commit the deal before any party voted to abort. A *proof of abort* proves that some party voted to abort before every party voted to commit. A party can rescind an earlier commit vote by voting to abort (for example, if the deal is taking too long to complete). To ensure strong liveness, once a compliant party has voted to commit, it must wait long enough to give the other parties a chance to vote before it changes its mind and votes to abort.

Recall that a commit protocol is *decentralized* if there is no single blockchain accessed by all parties in any execution (see Section 5.6.2). The CBC protocol is not decentralized in this sense, because the CBC itself is a centralized "whiteboard" shared by all parties. This loss of decentralization is inevitable: no protocol that tolerates periods of asynchrony can be decentralized. A complete formal proof is out of scope, but we outline an argument adapted from Fischer et al. [1985]. If all parties are compliant, then if the deal commits (resp. aborts) at any asset's blockchain, it must commit (resp. abort) at all of them. Initially, the deal's state is *bivalent*: both commit and abort are possible outcomes. But the deal's state cannot remain bivalent forever, so it must be possible to reach a (bivalent) *critical state* where each party is about to take a decisive step that will force the protocol to enter a *univalent* state where either a commit outcome or an abort outcome becomes inevitable. A potentially decisive step forcing an eventual commit cannot take place at a different blockchain than a potentially decisive step forcing an eventual abort, because then it would be impossible to determine which happened first, hence which one was truly decisive. It follows that in any such critical state, all parties must be about to call the same contract, violating decentralization.

5.7.1 RUNNING THE PROTOCOL

Here is how to execute the phases of a CBC protocol.

Clearing phase. The market-clearing service broadcasts a unique identifier D and a list of participating parties *plist* (this protocol does not require the t_0 starting time or Δ). One party records the start of the deal on the CBC by publishing an entry:

$$startDeal(D, plist).$$

The calling party must appear in the *plist*. If more than one *startDeal* for D is recorded on the CBC, the earliest is considered definitive.

Escrow phase.   Each party places its outgoing assets in escrow:

$$escrow(D, plist, h, a, \ldots).$$

Here, h is the hash that identifies a particular *startDeal* entry on the CBC that started the deal; it is needed in case there is more than one such entry on the CBC. The ellipsis indicates arguments that vary depending on the algorithm used to implement the CBC, as discussed in Section 5.7.4. As in the timelock protocol, the sender must be the owner of asset a and a member of *plist*.

Transfer phase.   Party P transfers an asset (or assets) a tentatively owned by P to party Q by sending

$$transfer(D, a, Q)$$

to the escrow contract on the asset's blockchain. P must be the owner of a and Q must be in the *plist*.

Validation phase.   As before, each party checks that its proposed payoff is acceptable and that assets are properly escrowed with the correct *plist* and h.

Commit phase.   Each party X publishes either a commit or abort vote for D on the CBC:

$$commit(D, h, X) \quad \text{or} \quad abort(D, h, X)$$

where D is the deal identifier, and h is the *startDeal*. As usual, each voter must be in the start-of-deal *plist*. After voting, parties monitor the CBC until there are enough votes to commit or abort the deal. As discussed below, each party then assembles a *proof* of commit or abort to unlock escrowed assets.

5.7.2   WHAT COULD POSSIBLY GO WRONG?

When things go wrong, the CBC protocol permits fewer outcomes than the timelock protocol, because all compliant parties agree on whether the deal committed or aborted. Nevertheless, because we cannot constrain the behavior of deviating parties, even this protocol cannot enforce the classical "all-or-nothing" property of atomic transactions. For example, if (deviating) Carol erroneously sends 1001 coins to Alice, instead of the 101 expected, and all parties vote to commit, then (compliant) Alice ends up with a commission of 901 coins, an outcome that is neither "all" nor "nothing," even for compliant parties. Of course, such an outcome is unlikely in practice, but such distinctions matter when reasoning about correctness. Both the timelock and CBC protocols satisfy the (informal) safety property that no compliant party can end up "worse off."

5.7.3 CORRECTNESS

The correctness of the CBC protocol is mostly self-evident: safety is satisfied because compliant parties agree on whether a deal commits or aborts. Weak liveness is satisfied because any compliant party whose assets are locked up for too long will eventually vote to abort, and strong liveness is satisfied in periods when the network is synchronous because every party votes to commit before any party votes to abort.

5.7.4 CROSS-CHAIN PROOFS

It is easy for (active) parties to ascertain whether a deal committed or aborted; it is not so easy for passive contracts, which cannot directly observe other blockchains, to do so.

A deal's *decisive vote* is the one that determines whether the deal commits or aborts. A straightforward approach is to present each contract with a subsequence of the CBC's blocks, starting with the deal's first *startDeal* record, and ending with its decisive vote. But how can the contract tell whether the blocks presented are really on the CBC? The answer depends partly on the kind of algorithm underlying the CBC blockchain.

5.7.5 BYZANTINE FAULT-TOLERANT CONSENSUS

Let us assume the CBC relies on *Byzantine fault-tolerant* (BFT) consensus [Abraham et al. 2018, Androulaki et al. 2018a, Castro and Liskov 1999, Tendermint 2015]. BFT protocols guarantee safety even when communication is asynchronous, and they ensure liveness when communication becomes synchronous after the global stabilization time (GST).

Blocks are approved by a known set of $3f + 1$ *validators*, of which at most f can deviate from the protocol. (The details of how validators reach consensus on new blocks are not important here.) To support long-term fault tolerance, the blockchain is periodically *reconfigured* by having at least $2f + 1$ current validators elect a new set of validators. For ease of exposition, assume each block contains the next block's group of validators and their keys.

Each block in a BFT blockchain is vouched for by a certificate containing at least $f + 1$ validator signatures of that block's hash. (Any $f + 1$ signatures are enough because at least one of them must come from an honest validator.) The sequence of blocks and their certificates can be used as a proof. The contract on the asset blockchain will be able to check this proof as long as it knows the first block's set of validators, accomplished by passing the $3f + 1$ initial block validators as an argument to each of the deal's escrow contracts (in place of the ellipses). Parties must identify correct validators when putting assets in escrow, and they must check validators' credentials before voting to commit.

Checking the proof as just described is a lot of work; the proof is likely to be spread over many blocks, each containing a large number of entries. Furthermore, we cannot shorten the proof by omitting irrelevant block entries, because then a malicious party might fool a contract into making a wrong decision. But there are many ways to make BFT proofs more efficient.

A straightforward optimization is to take advantage of the fact that the CBC has validators. This allows the parties to request certificates from the CBC. Such a certificate would vouch for the current state of the deal (active, committed, aborted). This certificate alone would constitute a proof provided the original validators are still active, and otherwise the party must also provide the chain of validators across each reconfiguration.

5.7.6 PROOF-OF-WORK (NAKAMOTO) CONSENSUS

Proofs of commit or abort generated by a CBC implemented using proof-of-work (PoW) consensus (like Bitcoin [Nakamoto 2009] or Ethereum [Ethereum 2021]) are possible, but care is needed because such blockchains lack *finality*: any proof might be contradicted by a later proof, although forging a later, contradictory proof becomes more expensive to the adversary the longer it waits. (Kiayias et al. [2016, 2017] propose changes to standard PoW protocols that would make such "proofs of proof of work" more compact.)

Here is a scenario where Alice can construct a fake "proof of abort" for a proof-of-work CBC. As soon as the deal execution starts, Alice (perhaps aided by partners in crime) privately mines a block that contains an *abort* vote from Alice. When her part of the deal is complete, however, Alice publicly sends a *commit* vote to the CBC. If, by the time all parties have voted *commit*, Alice was able to mine a private *abort* block, then Alice can use that fake proof of abort to halt outgoing transfers of her assets, while using the legitimate proof of commit to trigger incoming transfers.

In the spirit of proof of work, such an attack can be made more expensive by requiring a proof of commit or abort to include some number of *confirmation* blocks beyond the one containing the decisive vote, forcing Alice to outperform the rest of the CBC's miners for an extended duration. To deter rational cheaters, the number of confirmations required should vary depending on the value of the deal, implying that high-value deals would take longer to resolve than lower-value deals.

To summarize, while it is technically possible to produce commit or abort proofs from a proof-of-work CBC, the result is likely to be slow and complex. In the same way a proof-of-work blockchain can fork, a "proof of proof of work" [Kiayias et al. 2016] can be contradicted by a later "proof of proof of work." Similarly, to make the production of contradictory proofs expensive, the proof's difficulty must be adjusted to match the value of the assets transferred by the deal. By contrast, a BFT certificate of commit or abort is final, and independent of the value of the deal's assets.

5.8   RELATED WORK

As noted, in a *cross-chain swap* [bitcoinwiki 2018, Bowe and Hopwood 2018, Decred 2018, Herlihy 2018, Nolan 2018, Komodo Platform 2018, Zakhary et al. 2019, Zyskind et al. 2018], each party transfers an asset to another party and halts. Cross-chain swaps are attractive because they reduce or eliminate the use of exchanges, some of which have proved to be untrustworthy [Wikipedia 2019a, 2019b]. However, cross-chain swaps lack the power to express the simple ticket brokerage deal described in our example, as well as auctions and other conventional financial transactions.

To our knowledge, the only cross-chain swap protocols used in practice are *hashed time-locked contracts* [bitcoinwiki 2018, Bowe and Hopwood 2018, Decred 2018, Nolan 2018, Komodo Platform 2018]. Herlihy [2018] generalizes prior two-party cross-chain swap protocols to a protocol for multi-party swaps on arbitrary strongly connected directed graphs. Herlihy also observes that the classical "all-or-nothing" correctness property is ill-suited to cross-chain swaps, and proposes an alternative correctness property that is more specialized than the one presented here because it is formulated explicitly in terms of direct swaps, not the more general structures permitted by cross-chain deals. For example, Herlihy assumes that any swap outcome where a party receives only partial inputs and partial outputs is unacceptable, but the notions of correctness introduced here allow parties to specify whether some such partial deal outcomes are acceptable.

The timelock commit protocol presented here has a simpler structure than the one proposed by Herlihy. That protocol used secrets held by a carefully chosen subset of parties. Our protocol replaces secrets with votes performed by everyone, so it is possible to treat all parties uniformly, and there is no need for a careful contract deployment phase. Our protocol also clarifies when parties review the transactions' final outcomes. Both commit protocols use timeout mechanisms based on path signatures.

Zakhary et al. [2019] propose a cross-chain commitment protocol that does not use hashed timelocks. Instead, participating blockchains exchange "proofs" of state changes, somewhat similar to our CBC proposal, but because parties need to register their intended transfers at the start, this protocol supports only unconditional swaps, not full-fledged deals.

Off-chain payment networks [Decker and Wattenhofer 2015, Green and Miers 2016, Heilman et al. 2019, Poon and Dryja 2016, Raiden Network 2018] and state channels [Coleman et al. 2018] use hashed timelock contracts to circumvent the scalability limits of existing blockchains. They conduct repeated off-chain transactions, finalizing their net transactions in a single on-chain transaction. The use of hashed timelock contracts ensures that parties cannot be cheated if one party tries to settle an incorrect final state. Lind et al. [2017] propose using trusted execution environments in hardware to ease synchrony requirements. It remains to be seen whether off-chain networks can be applied to cross-chain deals. Arwen [Heilman et al. 2019] supports multiple off-chain atomic swaps between parties and exchanges, but their protocol is specialized to currency trading and does not seem to support non-fungible assets. Komodo [Komodo Platform 2018] supports off-chain cross-platform payments.

Sharded blockchains [Al-Bassam et al. 2017, Kokoris Kogias et al. 2018] address scalability limits of blockchains by partitioning the state into multiple shards so that transactions on different shards can proceed in parallel, and they support multi-step atomic transactions spanning multiple shards. An atomic transaction that spans multiple shards is executed at the client in Chainspace [Al-Bassam et al. 2017], or at the server in OmniLedger [Kokoris Kogias et al. 2018]. In these systems a transaction represents a single trusted party and there is no support for transactions involving untrusted parties.

Chainspace [Al-Bassam et al. 2017] allows transactions to specify immutable proof contracts to be executed at the server. The proofs are used to validate client execution traces resembling optimistic concurrency control. Channels [Androulaki et al. 2018b], an extension of OmniLedger Atomix protocols, uses proofs in a two-phase protocol similar to our CBC, for atomic untrusted cross-shard single-step multi-party UTXO (unspent transaction output) transfers [Investopedia 2019], but does not support multi-step deals or non-fungible assets.

The BAR (Byzantine, altruistic, and rational) computation model [Aiyer et al. 2005, Clement et al. 2008] supports cooperative services spanning autonomous administrative domains that are resilient to Byzantine and rational manipulations. Like Byzantine fault-tolerant systems, BAR-tolerant systems assume a bounded number of Byzantine faults, and as such do not fit the adversarial deal model, where any number of parties may be Byzantine.

The CBC somewhat resembles an *oracle* [Peterson et al. 2018], a trusted data feed that reports physical-world occurrences to contracts.

An early precursor of adversarial commerce was the study of *federated databases* [Sheth and Larson 1990], which addressed the problem of coordinating and committing transactions that span multiple autonomous, mutually untrusting, heterogeneous data stores. (Federated databases did not attempt to tolerate arbitrary Byzantine behavior.)

5.9 ACKNOWLEDGMENTS

Maurice Herlihy was supported by NSF grant 1917990.

BIBLIOGRAPHY

I. Abraham, G. Gueta, and D. Malkhi. Hot-Stuff: The linear, optimal-resilience, one-message BFT devil. In *Proceedings of PODC 2019, CoRR*, abs/1803.05069, 2018. 135, 148

A. S. Aiyer, L. Alvisi, A. Clement, M. Dahlin, J.-P. Martin, and C. Porth. BAR fault tolerance for cooperative services. In *Proceedings of the 20th ACM Symposium on Operating Systems Principles, SOSP '05, New York, New York, USA*, pages 45–58. ACM, 2005. 151

M. Al-Bassam, A. Sonnino, S. Bano, D. Hrycyszyn, and G. Danezis. Chainspace: A sharded smart contracts platform, 2017. *CoRR*, abs/1708.03778. 151

E. Androulaki, A. Barger, V. Bortnikov, C. Cachin, K. Christidis, A. De Caro, D. Enyeart, C. Ferris, G. Laventman, Y. Manevich, S. Muralidharan, C. Murthy, B. Nguyen, M. Sethi, G. Singh, K. Smith, A. Sorniotti, C. Stathakopoulou, M. Vukolić, S. W. Cocco, and J. Yellick. Hyperledger Fabric: A distributed operating system for permissioned blockchains. In *Proceedings of the 13th EuroSys Conference, EuroSys '18, New York, New York, USA*, pages 30:1–30:15. ACM, 2018a. 148

E. Androulaki, C. Cachin, A. D. Caro, and E. Kokoris-Kogias. Channels: Horizontal scaling and confidentiality on permissioned blockchains. In *ESORICS, 23rd European Symposium on Research in Computer Security*, volume 11098 of *Lecture Notes in Computer Science*, pages 111–131. Cham, Springer, 2018b. 151

P. A. Bernstein, V. Hadzilacos, and N. Goodman. *Concurrency Control and Recovery in Database Systems*. Addison-Wesley Longman Publishing Co., Inc., 1986. 138, 146

bitcoinwiki. Atomic cross-chain trading, 2018. https://en.bitcoin.it/wiki/Atomic_cross-chain_trading. (Accessed 9 Jan. 2018.) 150

S. Bowe and D. Hopwood. Hashed time-locked contract transactions, 2018. https://github.com/bitcoin/bips/blob/master/bip-0199.mediawiki. (Accessed 9 Jan. 2018.) 150

M. Castro and B. Liskov. Practical Byzantine fault tolerance. In *Proceedings of the 3rd Symposium on Operating Systems Design and Implementation, OSDI '99, Berkeley, California, USA*, pages 173–186. USENIX Association, 1999. 148

J. Chester. Your guide on Bitcoin's Lightning Network: The opportunities and the issues, June 2018. https://www.forbes.com/sites/jonathanchester/2018/06/18/your-guide-on-the-lightning-network-the-opportunities-and-the-issues/#6c8d8c0f3677N. (Accessed 11 Dec. 2018.) 145

A. Clement, H. Li, J. Napper, J. P. M. Martin, L. Alvisi, and M. Dahlin. BAR primer. In *Proceedings of the International Conference on Dependable Systems and Networks (DSN), DCC Symposium*, pages 287–296. IEEE, 2008. 151

J. Coleman, L. Horne, and L. Xuanji. Counterfactual: Generalized state channels, 2018. http://l4.ventures/papers/statechannels.pdf. 150

C. Decker and R. Wattenhofer. A fast and scalable payment network with Bitcoin duplex micropayment channels. In A. Pelc and A. A. Schwarzmann, editors, *Stabilization, Safety, and Security of Distributed Systems*, pages 3–18. Springer International Publishing, 2015. 150

Decred. Decred cross-chain atomic swapping, 2018. https://github.com/decred/atomicswap. (Accessed 8 Jan. 2018.) 150

C. Dwork, N. Lynch, and L. Stockmeyer. Consensus in the presence of partial synchrony. *J. ACM* 35(2):288–323, April 1988. 145

Ethereum. https://github.com/ethereum/. (Accessed 6 July 2021.) 135, 149

M. J. Fischer, N. A. Lynch, and M. S. Paterson. Impossibility of distributed consensus with one faulty process. *J. ACM* 32(2):374–382, April 1985. 140, 146

Y. Gilad, R. Hemo, S. Micali, G. Vlachos, and N. Zeldovich. Algorand: Scaling Byzantine agreements for cryptocurrencies. In *Proceedings of the 26th Symposium on Operating Systems Principles, SOSP '17, New York, New York, USA*, pages 51–68. ACM, 2017. 135

M. Green and I. Miers. Bolt: Anonymous payment channels for decentralized currencies. Cryptology ePrint Archive, Report 2016/701, 2016. https://eprint.iacr.org/2016/701. 150

E. Heilman, S. Lipmann, and S. Goldberg. The Arwen trading protocols, January 2019. https://www.arwen.io/whitepaper.pdf. (Accessed 23 Feb. 2019.) 150

M. Herlihy. Atomic cross-chain swaps. In *Proceedings of the 2018 ACM Symposium on Principles of Distributed Computing, PODC '18, New York, New York, USA*, pages 245–254. ACM, 2018. 144, 150

Investopedia. UTXO, 2019. https://www.investopedia.com/terms/u/utxo.asp. (Accessed 7 Apr. 2019.) 151

A. Kiayias, N. Lamprou, and A.-P. Stouka. Proofs of proofs of work with sublinear complexity. In *International Conference on Financial Cryptography and Data Security*, pages 61–78, 2016. 149

A. Kiayias, A. Miller, and D. Zindros. Non-interactive proofs of proof-of-work. Cryptology ePrint Archive, Report 2017/963, 2017. https://eprint.iacr.org/2017/963. 149

E. Kokoris Kogias, P. S. Jovanovic, L. Gasser, N. Gailly, E. Syta, and B. A. Ford. OmniLedger: A secure, scale-out, decentralized ledger via sharding. In *2018 IEEE Symposium on Security and Privacy (SP)*, page 16, 2018. 151

Komodo Platform. The BarterDEX whitepaper: A decentralized, open-source cryptocurrency exchange, powered by atomic-swap technology. https://supernet.org/en/technology/whitepapers/BarterDEX-Whitepaper-v0.4.pdf. (Accessed 9 Jan. 2018.) 150, 151

Libra Association. An introduction to Libra, 2019. https://libra.org/en-US/wp-content/uploads/sites/23/2019/06/LibraWhitePaper_en_US.pdf. (Accessed 24 Sept. 2019.) 135

J. Lind, I. Eyal, F. Kelbert, O. Naor, P. R. Pietzuch, and E. G. Sirer. Teechain: Scalable blockchain payments using trusted execution environments. *CoRR*, abs/1707.05454, 2017. 150

A. Miller, Y. Xia, K. Croman, E. Shi, and D. Song. The honey badger of BFT protocols. In *Proceedings of the 2016 ACM SIGSAC Conference on Computer and Communications Security, CCS '16, New York, New York, USA*, pages 31–42. ACM, 2016. 136

S. Nakamoto. Bitcoin: A peer-to-peer electronic cash system, May 2009. https://bitcoin.org/bitcoin.pdf. 135, 149

T. Nolan. Atomic swaps using cut and choose. https://bitcointalk.org/index.php?topic=1364951. (Accessed 9 Jan. 2018.) 150

J. Peterson, J. Krug, M. Zoltu, A. K. Williams, and S. Alexander. Augur: A decentralized oracle and prediction market platform. https://www.augur.net/whitepaper.pdf. (Accessed 7 Apr. 2019.) 151

J. Poon and T. Dryja. The Bitcoin Lightning Network: Scalable off-chain instant payments, Jan. 2017. https://lightning.network/lightning-network-paper.pdf. (Accessed 29 Dec. 2017.) 145, 150

Raiden Network. What is the Raiden Network? https://raiden.network/101.html. (Accessed 26 Jan. 2018.) 150

A. P. Sheth and J. A. Larson. Federated database systems for managing distributed, heterogeneous, and autonomous databases. *ACM Comput. Surv.* 22(3), September 1990. 151

Tendermint, October 2015. http:/https://github.com/tendermint/tendermint/wiki. Commit c318a227.
148

Wikipedia. Mt. Gox, 2019a. https://en.wikipedia.org/wiki/Mt._Gox. (Accessed 6 Apr. 2019.) 150

Wikipedia. Quadriga Fintech Solutions, 2019b. https://en.wikipedia.org/wiki/Quadriga_Fintech_
Solutions. (Accessed 6 Apr. 2019.) 150

V. Zakhary, D. Agrawal, and A. El Abbadi. Atomic commitment across blockchains. *CoRR*,
abs/1905.02847, 2019. 150

G. Zyskind, C. Kisagun, and C. FromKnecht. Enigma Catalyst: A machine-based investing platform and
infrastructure for crypto-assets, 2018. https://www.enigma.co/enigma_catalyst.pdf. (Accessed 25
Jan. 2018.) 150

AUTHORS' BIOGRAPHIES

Maurice Herlihy has an AB in Mathematics from Harvard University, and a PhD in Computer
Science from MIT. He is currently the An Wang Professor of Computer Science at Brown
University. He has served on the faculty of Carnegie Mellon University and the staff of DEC
Cambridge Research Lab. He is the recipient of the 2003 Dijkstra Prize in Distributed Computing, the 2004 Gödel Prize in theoretical computer science, the 2008 ISCA influential paper
award, the 2012 Edsger W. Dijkstra Prize, and the 2013 Wallace McDowell award. He received
a 2012 Fulbright Distinguished Chair in the Natural Sciences and Engineering Lecturing Fellowship, and he is a fellow of the ACM, as well as a fellow of the National Academy of Inventors,
the National Academy of Engineering, and the National Academy of Arts and Sciences.

Barbara Liskov is an Institute Professor at MIT. Her research interests include distributed and
parallel systems, programming methodology, and programming languages. Liskov is a member
of the National Academy of Engineering, the National Academy of Sciences, the National
Inventors Hall of Fame, and the Massachusetts Academy of Sciences. She is a fellow of the
American Academy of Arts and Sciences and the Association for Computing Machinery, and a
charter fellow of the National Academy of Inventors. She received the ACM Turing Award in
2009, the IEEE Von Neumann medal in 2004, the IEEE Pioneer Award in 2018, a lifetime
achievement award from the Society of Women Engineers in 1996, the ACM SIGPLAN
Programming Language Achievement Award in 2008, the ACM SIGOPS Hall of Fame award
in 2012, and the Stanford Hero of Engineering award in 2019.

Liuba Shrira is a professor of computer science at Brandeis University. Her research interests
primarily involve distributed systems. She received her PhD from the Technion. She is also
affiliated with the MIT Computer Science and Artificial Intelligence Laboratory.

CHAPTER 6

Strategic Interactions in Blockchain: A Survey of Game-Theoretic Approaches

Bruno Biais, *HEC Paris and Toulouse School of Economics, Université Toulouse Capitole, France*
Christophe Bisière, *Toulouse School of Economics, Université Toulouse Capitole, France*
Matthieu Bouvard, *Toulouse School of Economics, Université Toulouse Capitole, France*
Catherine Casamatta, *Toulouse School of Economics, Université Toulouse Capitole, France*

6.1   INTRODUCTION

Blockchains are distributed ledgers maintained through a combination of technologies (e.g., cryptography, peer-to-peer networks) and protocols ensuring that nodes in the network come to an agreement on the current state of the ledger. To study the performance and reliability of these protocols, computer scientists traditionally draw a distinction between processes that conform to the protocol and faulty processes that don't conform to the protocol. For example, under proof of work (PoW), processes are miners, the protocol defined in Nakamoto [2008] considers the longest chain as representing the consensus, and faulty processes are attackers whose aim is to perturb consensus building by deviating from this longest chain rule. In this approach, the question of interest is the extent to which malicious nodes can succeed in breaking consensus, or equivalently, what proportion of honest nodes is required to maintain consensus. While this approach provides useful notions of robustness, it is silent on the reasons why processes adhere to the protocol or deviate from it.

This blind spot suggests a role for economic analysis to complement the work of computer scientists: economics offers a conceptual framework to model and understand the role of incentives that can be applied to blockchains. Instead of assuming specific behaviors (e.g., honest or faulty), the natural economic approach is to model processes as rational agents who choose actions to maximize their expected utility. This approach reflects the view that human decisions ultimately drive strategies in the blockchain and in particular whether processes conform to the protocol.

In addition, the economists' toolkit is able to account for situations where agents' strategies exhibit complex dependencies. In particular, economists draw a distinction between competitive and strategic behaviors. Competitive agents take the characteristics of the economic environment (e.g., prices) as given, and react optimally to it. Strategic players take into account the impact of their actions on the outcome of their interactions with other players. Game theory offers a framework to study strategic players and its application to blockchains is the focus of this survey.[1] Game theory is particularly well suited to the analysis of the agents in charge of validating transactions in a blockchain—for example, the miners under PoW. Indeed, in permissionless blockchains like Bitcoin or Ethereum, miners operate within large pools that coordinate their behavior and take into account the impact of their decisions on market share and profitability. Analyses of competitive behaviors in the blockchain are out of the scope of this chapter, and so are monetary economic analyses of cryptocurrencies. Halaburda and Haeringer [2019] offer a broad survey of blockchains and cryptocurrencies.

Bringing game theory into the analysis of blockchains helps shed light on miners' choices and how they are affected by the blockchain environment. On the one hand, miners—or, more generally, validators—face complex decision problems. For instance, they need to decide to which existing block to chain new blocks of transactions, anticipating where other miners will chain their own blocks. We discuss which equilibrium strategies emerge in this type of game and whether they are compatible with the integrity and reliability of the blockchain. On the other hand, the structure of these games and the resulting equilibrium strategies are affected by multiple aspects of the blockchain design and environment. We discuss the impact of: (i) transaction fees, (ii) the formation of mining pools, (iii) investments in computing capacity, and (iv) alternative consensus protocols like proof of stake (PoS).[2]

We acknowledge that our selection of papers is partial, and does not do justice to the rich and growing literature on these issues (see Liu et al. [2019] for a comprehensive survey of game-theoretical approaches to the blockchain). Our objective is not to be comprehensive, but to focus on some key economic mechanisms induced by the blockchain protocol.

6.2   MINING STRATEGIES

In a blockchain, the history of transactions is represented by a chain of transaction blocks. The goal of a consensus protocol is to ensure that participants agree on which chain represents this history.

---

[1] We focus on non-cooperative games. This book's Chapter 7 by Marianna Belotti and Stefano Moretti covers cooperative games.

[2] While this survey focuses on strategic interactions between the participants of the blockchain, Badertscher et al. [2018] consider the game between a protocol designer and an attacker.

Under proof of work, this is implemented through a distributed lottery: a miner is selected to append a block to the blockchain if it is the first to solve a numerical problem by random trials—an activity called "mining." Consensus is achieved when there is a single chain, without forks. This obtains when miners follow the "longest chain rule" suggested by Nakamoto [2008]. Will miners follow this suggestion in equilibrium? Or should we expect them to deviate from the longest chain rule? In that case, which patterns can emerge? Will forks be transient or persistent? What economic forces will make them more or less likely? The papers surveyed in this section address these issues.

6.2.1   THE LONGEST CHAIN RULE AND FORKS

One of the first papers to consider consensus formation in the blockchain PoW protocol as the outcome of a game is Kroll et al. [2013]. The paper makes important observations. First, it views a miner's strategy as a mapping from "the blockchain structure [. . .] to a choice of which branch to mine on, that is, which block will be the parent of the newly mined block if the player wins." Second, it notes that "reward is only valuable if the newly mined block ends up on the long-term consensus chain." In this context, the paper states that following the longest chain rule is a Nash equilibrium strategy, but that there are other equilibria. While Kroll et al. [2013] thus provide important economic insights, they do not formally describe the extensive form of the game, nor do they provide an explicit characterization of the Nash equilibria.

The work in Biais et al. [2019a] and Kiayias et al. [2016] formally analyzes interaction between miners as a stochastic game. Both papers consider miners choosing to which parent block in the current state of the blockchain to chain the block they currently mine. In Biais et al. [2019a], for all parameter values, and in particular for any distribution of computing power among miners, following the longest chain rule is a Nash equilibrium strategy. In contrast, in Kiayias et al. [2016], this is the case when each miner's computing power is sufficiently small. The two papers make different assumptions about the extensive form of the mining game, highlighting different economic forces at play.

In line with Kroll et al. [2013], in Biais et al. [2019a] the value of a miner's reward increases as the branch to which his block belongs gets closer to consensus. Specifically, Biais et al. [2019a] assume that the reward for a block is larger when more miners chain their blocks to its branch. This assumption is made to capture the idea that rewards are paid in units of cryptocurrency, whose value is higher when more agents accept it.[3] An important consequence of this assumption is that miners want to chain their block to the branch they expect others to adopt—that is, mining in the

---

[3] For analyses of the value of cryptocurrencies, see, for example, Athey et al. [2016], Biais et al. [2019c], Cong et al. [2021b], Fanti et al. [2019], Fernández-Villaverde and Sanchez [2018], Gandal and Halaburda [2016], Garratt and Wallace [2018], Pagnotta [2018], Saleh [2018], Schilling and Uhlig [2019], or Zhu and Hendry [2019].

blockchain is a coordination game. Therefore, if miners anticipate that the other miners will follow the longest chain rule, their best response is to do the same. This explains why, in Biais et al. [2019a], following the longest chain rule is always a Nash equilibrium.

Kiayias et al. [2016] make a different assumption about rewards: "at every level, only one node is paid for, the first one which succeeds in having a descendant d generations later. [. . .] When this happens, every sibling (as well as its descendants)" gets no reward. In this framework a fork can only generate a reward if it reaches the d-block threshold before the honest branch. When miners' computing power is small, they have little chance to win that race if forking. Thus, in Kiayias et al. [2016], the longest chain rule is a Nash equilibrium if each miner's computing power is sufficiently small.

While both Biais et al. [2019a] and Kiayias et al. [2016] state assumptions under which the longest chain rule can be an equilibrium, Biais et al. [2019a] highlight the fact that coordination effects give rise to other (multiple) equilibria. Indeed, if a miner anticipates that the others will fork and abandon the longest chain, the best response is to do the same. This generates orphaned branches at equilibrium. Biais et al. [2019a] characterize equilibria in which such forks can be of arbitrary length and shows that equilibria with forks exist for any distribution of computing power among miners. Next, they investigate the strategic consequences of the k-block rule, which prevents miners from spending their rewards before k blocks are chained to the block they solved.[4] This rule generates vested interests, in the sense that miners who solved many blocks on a branch strongly prefer that this branch survives. Biais et al. [2019a] give conditions on parameter values such that the combination of vested interests and coordination effects gives rise to equilibria with persistent forks—that is, situations in which several competing branches grow simultaneously. An example of persistent forks is offered by the split between Ethereum and Ethereum Classic in July 2016, which led to two blockchains that still coexist today.

As discussed above, the way in which miners are compensated affects the nature of their strategic interaction. Kroll et al. [2013], Biais et al. [2019a], and Kiayias et al. [2016] analyze situations in which miners earn rewards from coinbase transactions. Miners, however, are not only compensated with block rewards, but also with transaction fees. While these fees are currently a small fraction of miners' rewards, they will become important when coinbase transactions disappear. To provide insights into this evolution, Carlsten et al. [2016] study how strategic interactions between miners are affected by transaction fees. With transaction fees, an important additional choice variable is which transactions and associated fees to include in one's block. Carlsten et al. [2016] further show that when a block includes transactions with large fees, miners have an incentive to fork and create an alternative block including some of these transactions. When doing so, they choose to leave some

---

[4] For Bitcoin, k is 100 blocks.

transactions out of their block, to induce subsequent miners to chain their own block to the fork. Carlsten et al. [2016] thus show that the strategic choice of transactions in order to earn large fees can give rise to equilibrium forks and protocol instability.[5]

6.2.2   DOUBLE SPENDING

The above papers point to coordination effects, vested interests, and transaction fees as potential causes of forks. Miners could also want to deviate from the longest chain rule in order to double spend. When other miners follow the longest chain rule, a deviating miner can successfully double spend if he is able to add sufficiently many blocks to create a chain longer than the original one. This, however, requires large computing capacity. Bonneau et al. [2016] analyze "bribery attacks" in which miners can obtain large computing capacity for a limited period by renting it. Teutsch et al. [2016] consider an alternative way to increase one's share of total computing power dedicated to mining: the attacker offers prizes to other miners for solving puzzles outside the blockchain, thus reducing the pace at which blocks are added on the public branch. By doing so, an attacker with sufficient initial capital can win the race—that is, ensure the attacker's own private chain is longer than the public one. While, like Kiayias et al. [2016], Teutsch et al. [2016] model forking as a race, Biais et al. [2019a] highlight the way that coordination effects also condition the success of double-spending attacks: an attack is successful if all participants anticipate so. Biais et al. [2019a] also derive explicit conditions under which double-spending attacks relying on coordination effects can be triggered. An important result of both Teutsch et al. [2016] and Biais et al. [2019a] is that double spending can be successful in equilibrium even if the double spender does not have the majority of computing power. Early analyses of blockchains pointed to attacks relying on 51% computing power as the major threat to protocol security. Game-theoretical approaches, however, show that consensus can be unstable even if no miner (or pool) has the majority of the computing power.

6.2.3   UPGRADES

Consensus is particularly difficult to achieve, and the risk of forks is particularly large, when decisions about the protocol must be made by participants. Indeed, in practice, most forks have been triggered by protocol upgrades, as illustrated for Bitcoin in Figure 6.1 (borrowed from Biais et al. [2019b]). Authors in Biais et al. [2019b] highlight the crucial role played by coordination effects in the emergence of these forks: If each miner anticipates the upgrade to be adopted (resp. rejected) by all the others, then the upgrade is adopted (resp. rejected) in equilibrium, irrespective of whether

---

[5] Houy [2016] offers an alternative analysis of the choice of which transactions to include in a block, resulting from miners' trade-off between the probability that a solved block is accepted by others and the revenue from transaction fees included in the block.

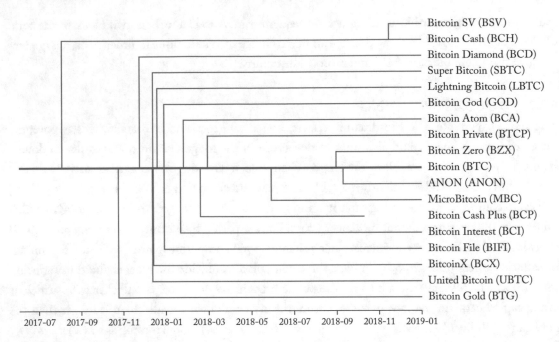

FIGURE 6.1: Bitcoin forks since 2017. The figure only depicts forks for which market data are available. Each branch starts at the time of the fork, and lasts as long as market data are available.

it is socially optimal or not. Moreover, Biais et al. [2019b] give conditions under which, if some miners derive private benefits from using one version of the protocol, equilibria with persistent forks (in which one branch accepts the upgrade while the other sticks to the previous version of the protocol) can be sustained. Barrera and Hurder [2018] study whether governance mechanisms can solve coordination problems in this context and show that two common voting schemes (majority rule and quadratic voting) can fail to eliminate suboptimal forks.

6.2.4   BLOCK WITHHOLDING STRATEGIES

The previous section considered analyses in which blocks were published as soon as they were solved. Miners, however, could choose to withhold some blocks, a strategy referred to as selfish mining. Eyal and Sirer [2014] offer the first analysis of this issue from a game-theoretical perspective. The paper shows that if a colluding group of miners follows a selfish mining strategy, while the others are honest (i.e., stick to the longest chain rule), then the colluding group of miners obtains a fraction of total rewards that is larger than its fraction of the computing power, and consequently honest miners obtain a fraction of total rewards smaller than their share of computing power. This occurs even if the group of colluding miners does not control the majority of the computing power.

This can be seen by considering the following simple version of the selfish mining strategy: in the initial stage of the strategy, the selfish miner finds a block and withholds it until the next block is discovered. If the next block is found by an honest miner, the selfish miner proposes a branch of the same length as that of the honest miner, and stands a chance to attract consensus. If the next block is found by the selfish miner, his branch is the longest one and therefore attracts consensus. While this strategy reduces the expected reward of the selfish miner relative to the situation in which all miners follow the longest chain rule, it reduces even more the expected reward of the honest miners. As noted by Eyal and Sirer [2014]: "The key insight behind the selfish mining strategy is to force the honest miners into performing wasted computations on the stale public branch."

Under Eyal and Sirer's [2014] assumption that the attacker's objective is to maximize his share of total revenue, selfish mining is a best response to honest mining because some blocks solved by honest miners become stale.[6] It is not a best response, however, if the attacker's objective is to maximize his expected reward. To motivate miners' focus on their share of total revenue, one could consider a setting in which that total revenue is constant.[7] In this setting, a miner's expected profit is reduced each time another miner obtains a reward. To maximize his profit, each miner should then maximize his share of total rewards. Selfish mining can also be profitable if it leads to a reduction in difficulty after some blocks are orphaned, as analyzed by Grunspan and Pérez-Marco [2019].

In practice, to the best of our knowledge, there is no compelling evidence that selfish mining is prevalent. This may be due to the conceptual difficulties to rationalize selfish mining discussed above.

6.3   THE SUPPLY OF MINING SERVICES

Blockchain is designed to operate as an open network, in which entry is free. An important question is therefore whether miners' decentralized entry and capacity decisions are socially efficient. This section surveys papers studying the determinants and organization of the supply of mining services.

6.3.1   COMPUTING CAPACITY CHOICES

A key feature of PoW protocols is that the difficulty of the hash puzzle adjusts to keep the average time between two blocks constant. Thus, when a miner increases her computing capacity, the difficulty of the hash puzzle increases for all participants. In this context, the probability that a

---

[6] Sapirshtein et al. [2016] extend the analysis of Eyal and Sirer [2014] by considering a larger set of selfish mining strategies, while retaining the assumption that the attacker maximizes his share of total revenue.

[7] For example, to shed light on selfish mining, Beccuti and Jaag [2017] analyze a setting in which there are only two fixed rewards, the allocation of which depends on the miners' reporting strategies.

miner solves a hash puzzle and obtains a reward is determined by her computing capacity *relative* to the total computing capacity on the blockchain.

Dimitri [2017] studies a simultaneous game among n miners who choose how much computing capacity to install. When choosing capacity, each miner takes into account its impact on her cost, as well as on her probability to solve a block, given the update in protocol difficulty. Intuitively, computing capacity plays a similar role to quantity in a Cournot game—that is, strategic miners choose to limit their impact on difficulty to maximize profits. This results in strictly positive equilibrium profits for miners, and implies that several miners are simultaneously active in equilibrium.

Biais et al. [2019a], however, point out that, in the above game, each miner exerts a negative externality on the others when increasing her own computing capacity. Indeed, a miner that builds up capacity makes it more difficult for the other miners to collect block rewards. Because of this negative externality, the computing capacity investment game can be interpreted as an arms race. Indeed, Biais et al. [2019a] show that in this context equilibrium computing power is inefficiently high, relative to the social optimum.

Arnosti and Weinberg [2018] also consider a model in which miners invest in computing capacity to increase their probability to earn a fixed reward. They show that when miners' costs are not symmetric, miners with lower marginal costs end up with a disproportionately higher share of total computing power. Thus, mild cost asymmetries result in highly concentrated mining power.

Alsabah and Capponi [2019] extend the work in Dimitri [2017], Arnosti and Weinberg [2018], and Biais et al. [2019a] to include an initial R&D stage, during which miners invest in research to develop better hashing technologies. Alsabah and Capponi [2019] show that, as with the computing capacity acquisition game in Dimitri [2017], Arnosti and Weinberg [2018], and Biais et al. [2019a], the R&D game is an arms race. In this context, limitations of property rights on research output, such as spillovers or lack of non-compete clauses, can improve welfare. Similar to Arnosti and Weinberg [2018], Alsabah and Capponi [2019] point out that the R&D game among miners can result in the centralization of the mining industry.

In Dimitri [2017] and Biais et al. [2019a], miners make positive profit when the number of miners n is finite. But in the framework of these models, when n goes to infinity, individual profits go to zero. A similar result obtains in the free entry game analyzed by Ma et al. [2018].

Dimitri [2017] and Biais et al. [2019a] consider a static computing capacity choice game. In contrast, Ma et al. [2018] and Prat and Walter [2018] focus on the dynamics of computing capacity. One could expect that as the dollar-value of Bitcoin rises and rewards from mining increase, entry and capacity acquisition should occur. Figure 6.2 shows that there is indeed some correlation between hashrate and Bitcoin price, but that the relation between the two variables is complex. To analyze this relation, Prat and Walter [2018] incorporate two key features of the environment: First, investment in computing capacity is largely irreversible. Second, the dollar-value of Bitcoin is highly volatile. Thus, when deciding to increase capacity, miners face a real option problem. The optimal policy

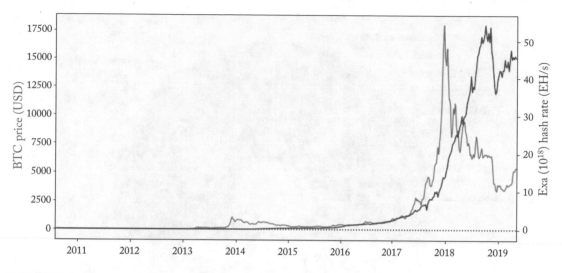

FIGURE 6.2: Evolution of hash rate and Bitcoin price. (Source: Authors' own computations and www.blockchain.com.)

entails a threshold instantaneous revenue from mining that triggers new investment. This barrier is reflective because as soon as miners invest, the difficulty adjustment pushes their revenue down. Calibrating the model generates a prediction of miners' investment in computing capacity from the time series of dollar/Bitcoin exchange rates. Prat and Walter [2018] show that this prediction closely matches the empirical evolution of installed capacity on the Bitcoin network.

In the previous models, an increase in the price of the cryptocurrency makes block rewards more valuable, which in turn stimulates investments in computing capacity. Pagnotta [2018] introduces a feedback loop where investment in computing capacity makes the blockchain more secure, which stimulates users' demand for the cryptocurrency and pushes its price up. This loop allows for the co-existence of multiple self-fulfilling equilibria. In one, because the cryptocurrency price is zero, no miner is active, which makes the blockchain insecure and users unwilling to pay any strictly positive price for using the cryptocurrency to transact. On the other hand, an equilibrium with strictly positive prices, active miners, and a positive demand from users may also be sustained. In a dynamic, overlapping-generation version of this model, Pagnotta [2018] links the installed capacity on the blockchain to the probability that the value of the cryptocurrency collapses to 0 in the next period, and further shows that these feedback effects may amplify volatility.

6.3.2 MINING POOLS

Given the increase in computing capacity depicted in Figure 6.2, any individual miner with limited computing capacity stands a very small chance of solving a block. Risk-averse miners, however,

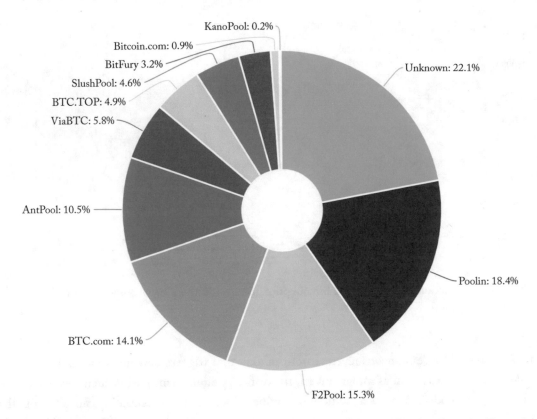

FIGURE 6.3: An estimation of hashrate distribution amongst the largest mining pools on 20 December 2019. (Source: www.blockchain.com.)

would benefit from mutualizing block discovery risk. Mining pools offer a vehicle for such risk sharing. Figure 6.3 graphically represents the distribution of computing power across mining pools. In this context, the following issues arise: Do mining pools provide efficient risk sharing? Do they exert market power? Can they adopt strategic behaviors that could undermine the functioning of the blockchain?

Cong et al. [2021a] analyze the provision of risk sharing services by pools and its consequences on miners' investment in computing capacity. Risk sharing works as follows: When miners decide to mine alone, they may obtain the full block reward, but with very low probability. By contrast, if they join a pool, they obtain a fraction of every block reward in proportion to their share of computing power.[8] When miners are risk averse, they prefer the latter to the former. Benefitting

---

[8] Rosenfeld [2011], Schrijvers et al. [2017], and Fisch et al. [2017] analyze the properties of different rewarding schemes.

from such insurance induces miners to acquire more computing power than they would if they were solo mining. In this context, Cong et al. [2021a] analyze how competing pools set their membership fees. Equilibrium depends on whether pools have captive members or not. First consider the case in which there are no captive miners. Pools compete à la Bertrand, driving equilibrium fees down to zero. If a pool was charging positive fees, no miner would choose to participate in it. Not participating in this pool would not reduce risk sharing since there is no one in this pool. Second, turn to the case in which pools have captive miners. A miner who considers participating in a pool compares the pool's fee to the benefit of sharing risk with the captive members. As long as the benefit is strictly positive, the pool can charge a strictly positive fee and still attract non-captive members. This effect is stronger for pools with a large captive base. Thus pools with a larger captive base charge higher fees, and yet remain larger than the others. Moreover, these large mining pools partially internalize the negative externality imposed by the computing power of their participants on the protocol difficulty. This further contributes to driving their fees up. It also tends to reduce their growth compared to that of smaller pools. This result alleviates concerns regarding large pools' ability to capture an excessive share of the market.

Ferreira et al. [2019], however, argue that the structure of the market for specialized mining equipment known as ASICs (application-specific integrated circuits) can foster mining pool concentration. For instance, Bitmain, the dominant ASIC producer, owns two particularly large mining pools, AntPool and BTC.com (see Figure 6.3). Ferreira et al. [2019] propose a model where ASICs are undifferentiated products and producers have a constant marginal cost. In the presence of sunk costs of production, there is no profitable entry once one producer is active, as Bertrand competition would drive the price to marginal cost and make it impossible to cover sunk costs. This leads to concentration in the ASIC production market, which then spills over into the mining pool market. The ASIC producer has an incentive to offer mining pool services with low fees to stimulate the entry of new miners and the demand for ASICs. Other pools do not have such incentives, so the mining fees of the ASIC producers are lower than those of the other pools. Because of lower fees, the producer-affiliated pool captures a large share of miners, and can be in a position to exert disproportionate control over the blockchain.

Beyond the exertion of market power, large mining pools could adopt strategic behaviors further threatening the blockchain, as analyzed in Eyal [2015] and Johnson et al. [2014]. Eyal [2015] studies whether pools should send their own miners to competing pools where they would share the rewards without contributing to block discovery. This can be achieved if infiltrating miners withhold their full proofs of work and yet obtain rewards from the infiltrated pool by disclosing partial proofs of work (next-to-valid solutions used by the pool to measure the work of a miner). As in Eyal and Sirer [2014], the objective of players is not to maximize their expected rewards. Rather, it involves one's share of total rewards. By sending its miners to the competing pool, a pool reduces its own fraction of total rewards but can capture a share of its opponent's fraction of total rewards. Eyal [2015] shows that in any Nash equilibrium of this game, pools send a strictly positive number of miners to infiltrate the other pools.

Johnson et al. [2014] study whether a pool should allocate resources to increase its computing power, or to launch a DDoS (distributed denial of service) attack on its competitor. Again, it is assumed that pools seek to maximize their fraction of total rewards. Johnson et al. [2014] show that for parameter constellations such that both pools would be better off increasing their computing power, there exists an equilibrium in which both launch a DDoS attack, which can be interpreted as a form of prisoner's dilemma.

6.4   TRANSACTION FEES

Blocks in the chain have fixed limited capacity. Consequently, when the volume of transactions is large, there is congestion. In order to have their transactions processed quickly, users can offer fees to miners. Thus, transaction fees and congestion should be correlated. Figure 6.4 suggests that large congestion tends to be followed by large fees. The following issues arise in this context: How are transaction fees set? Do they efficiently allocate priority among users? To what extent do fees mitigate congestion? Chiu and Koeppl [2019], Easley et al. [2019], and Huberman et al. [2018] offer contrasting insights on these questions.

Chiu and Koeppl [2019] study the interaction between high valuation buyers and low valuation sellers using the blockchain to settle transactions. After trading, buyers (resp. sellers) can be hit by a shock reducing (resp. increasing) their valuation for the asset. Users hit by a shock no longer benefit from trading and therefore would prefer to default. To do so, as in double spending, they try to create a fork in the blockchain, in order to cancel the initial transaction. Successful forking requires sufficiently large computational power relative to that installed by honest miners. In Chiu and Koeppl [2019], installed computational power is increasing in the fees offered by blockchain users. Therefore, large fees reduce the risk of forks and default. To ensure that investors are willing to pay large fees, however, block size must be limited. Thus, the analysis of Chiu and Koeppl [2019] contributes to the debate on the optimal size of blocks, suggesting that limited block size and congestion can enhance the reliability of the blockchain.

In practice, as can be seen in Figure 6.4, fees have been small and the largest part of miners' compensation comes from coinbase transactions. Extending the model in Chiu and Koeppl [2019] to include coinbase rewards, Chiu and Koeppl [2017] highlight the idea that setting the number of blocks before an agent can derive utility from his transaction (the "confirmation lag") is also useful to prevent double spending. In a general equilibrium context, Chiu and Koeppl [2017] further show that, to maximize social welfare, it is optimal to reward miners via currency creation only, rather than

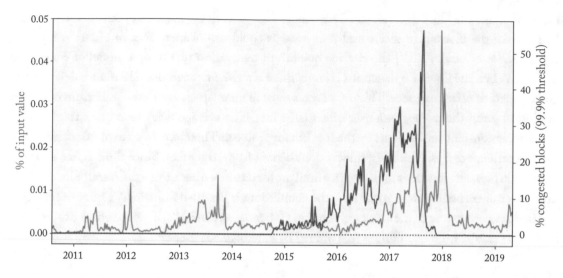

FIGURE 6.4: Evolution of transaction fees and congestion. (Source: Authors' own computations.) Transaction fees are expressed in percentage of Bitcoin transaction volume (input). Congestion is measured by the fraction of blocks for which transactions represent more than 99.9% of block capacity.

by using transaction fees. The reason is that currency creation is akin to a tax whose cost is spread across all potential buyers, while transaction fees are paid only by those buyers who actually enter into a transaction. Tsabary and Eyal [2018] highlight another drawback of relying on fees to incentivize miners. Transactions with attached fees arrive into the mempool through time. It follows that the value of solving a block increases through time as more transactions with fees can be incorporated into a block. As a result, if fees are relatively more important than block rewards, miners have an incentive to pause mining.

In Huberman et al. [2018], as in Chiu and Koeppl [2019], limited block size induces congestion, implying that users are willing to pay fees. In Huberman et al. [2018], users differ in their preference for immediacy and choose the fee they attach to transactions. Users with higher valuation for immediacy (i.e., more impatient users) find it optimal to offer larger fees. Thus, the most patient user attaches zero fee to his transaction and receives the lowest priority. Then, the user who is slightly more impatient optimally sets a fee f such that the most patient user is just indifferent between being last in the queue or being second to last and paying f. Iterating this logic yields the equilibrium fee schedule in which each user's fee equals the externality he imposes on other users by delaying their transactions. Hence, equilibrium is unique and priority allocation is efficient. The model further implies that miners' revenue from fees depends on congestion in the blockchain. When there are too few transactions, this revenue may be too low to attract enough computing power to make the blockchain secure. This echoes Chiu and Koeppl's [2019] point that congestion helps make the blockchain secure and that rewards could complement fees to attract computing capacity. Easley et al. [2019] also derive equilibrium transaction fees from a model of congestion pricing. When the flow of transactions is large, there is a risk of congestion. To reduce delays, users are tempted to offer large fees. This, however, results in a Pareto-dominated equilibrium outcome: hoping to jump the queue, each user offers large fees. The average delay to process transactions, however, is constant since it is set by the blockchain protocol. Therefore, in spite of offering fees, all users experience delays. Moreover, miners don't benefit from larger fees. Since there is free entry, an increase in fees only increases miners' entry until profits are driven back to zero. Overall, since neither users nor miners benefit from larger fees, the equilibrium is Pareto-dominated.[9] The corresponding inefficiency contrasts with the efficiency result in Huberman et al. [2018] in which efficient allocation of priority results from equilibrium fees offered by heterogenous users.[10]

6.5   ALTERNATIVE CONSENSUS PROTOCOLS

One drawback of PoW is the large electricity consumption and investment in hardware it entails. Proof of stake (PoS) is an alternative consensus mechanism that may avoid these deadweight losses. PoS implements a version of the following protocol. At regular time intervals, a validator is drawn from the pool of tokenholders and has the right to append a new block to an existing chain. Users with more tokens are more likely to be drawn; therefore agents with higher "stakes" exert more control over the state of the blockchain. The presumption is that these agents have more to lose if the blockchain malfunctions, and hence have better incentives to maintain consensus.[11] A related protocol is delegated proof of stake (DPoS), in which blockchain users vote for their preferred validator by placing their tokens on their candidate's name. As with PoS, users with more tokens indirectly exert more control through their votes.

Saleh [2021] formalizes this intuition. He revisits a problem known as "nothing at stake," thought to be a critical issue in PoS. Under PoW, mining a block on a given chain has an opportunity cost because the computing power devoted to that block cannot be deployed to mining a block on a different chain. By contrast, under PoS, adding a block is seemingly free for the validator who was drawn. This gives rise to the concern that validators would append any branch to ensure that

---

[9] The socially optimal level of fees needs not be zero, though, if a minimum number of miners is necessary to make the protocol viable. Easley et al. [2019] point out that equilibrium fees can be larger than this minimum, which decreases welfare.

[10] In Easley et al. [2019], users are homogenous and choose between an exogenous fee or no fee. As discussed in the paper, welfare analysis changes when users are heterogenous and can choose from a menu of fees.

[11] Fanti et al. [2019] study cryptocurrency market valuation in a PoS protocol. The protocol implies that block proposers can be requested to deposit tokens, which can be seized ("slashed") in case of misbehavior. It would be interesting to take a game-theoretic approach to study the strategic consequences of this feature of the protocol.

some of their blocks end up on the winning chain. This "nothing at stake" strategy would then perpetuate forks.[12] Saleh [2021] argues that this line of reasoning misses one cost for validators to indiscriminately add blocks, namely that it delays the time at which consensus is reached. The model starts with an exogenous fork, and assumes that tokens pay off only when the fork is resolved (i.e., when one branch becomes k blocks longer than the other branch) and only for those tokens that belong to the winning chain. Because agents discount future payoffs, the present value of a token during a fork is lower if the expected time at which the fork is resolved is further in the future. This gives an inherent incentive for validators with stakes to coordinate on appending only one chain to speed up consensus. As a result, the longest chain rule is an equilibrium, which implies in particular that a "nothing at stake" strategy is not a profitable deviation.[13]

While the above discussed protocols focus on the decision of the validator selected either by PoW or PoS, in other protocols (e.g., HoneyBadger, HotStuff, or Tendermint) a committee composed of a subset of deterministically selected processes executes an instance of practical Byzantine fault-tolerant consensus to decide on the next block to append. Amoussou-Guenou et al. [2019] analyze strategic interactions between validators in such protocols. They highlight the fact that coordination failures and free-riding within a committee can lead to equilibria for which the termination and validity properties do not hold. That is, the committee may fail to reach consensus or can end up accepting an invalid block. Such dysfunctional outcomes arise when rational committee members fail to check the validity of blocks or to send messages. Amoussou-Guenou et al. [2019] also show that there exists an equilibrium in which committee members are pivotal, which gives them the incentive to check validity and send messages, so that the termination and validity properties hold.

Manshaei et al. [2018] also take a game-theoretic approach to analyze multiple committees running in parallel to validate a non-intersecting set of transactions (a shard). They emphasize that rational agents can free-ride within a committee in which rewards are equally shared.

6.6   CONCLUSION

The analyses reviewed in this chapter shed light on two important sets of issues: (i) reliability of blockchain and consensus, and (ii) cost of blockchains.

Regarding the first issues, game-theoretical analyses underscore the principle that rational agents cannot be expected to blindly follow prescribed behavior, even if they do not derive any

---

[12] As written by Vitalik Buterin: "However, with the naive proof of stake algorithm described above, there is one serious problem: as some Bitcoin developers describe it, 'there is nothing at stake.' [. . . The] optimal strategy is to mine on any fork that you can find." Vitalik Buterin, 2014, https://blog.ethereum.org/2014/07/05/stake/.

[13] Brown-Cohen et al. [2018] discuss how the implementation of the quasi-random draw of a coin across all participants in the network opens scope for malicious strategies. In particular, the fact that the coin to be drawn several blocks ahead could be predictable increases the profitability of double spending and selfish mining.

private benefit from failing the blockchain. Rational, self-interested behavior can threaten blockchain stability and consensus for two types of reasons. On the one hand, coordination failures can generate forks. On the other hand, profit-maximizing agents can engage in manipulative behaviors, such as selfish mining or infiltrating pools. An important insight of game-theoretical approaches is that consensus can be unstable even if no miner or pool has the majority of computing power.

Regarding the cost of blockchains, two types of issues arise: Are transaction fees efficiently set? Is computing capacity optimal? Given a fixed maximum block size, fees can serve as useful price signals, either to incentivize investment in computing capacity or to allocate priority. Game-theoretical analyses, however, suggest that it would be efficient to relax block size constraint, and to rely on other features of the protocol to induce sufficient participation of miners. Another source of inefficiency is the negative externality imposed by miners on others when they increase their own computing capacity. This leads to an arms race with inefficiently high capacity. This underscores the need for more sober protocols than PoW, such as PoS. Besides the technical issues raised by the implementation of such protocols, it is likely that strategic interactions in these new environments will raise new challenges that the literature has only started to investigate.

BIBLIOGRAPHY

H. Alsabah and A. Capponi. Pitfalls of Bitcoin's proof-of-work: R&D arms race and mining centralization, 2019. Available at SSRN: http://dx.doi.org/10.2139/ssrn.3273982. 162

Y. Amoussou-Guenou, B. Biais, M. Potop-Butucaru, and S. Tucci-Piergiovanni. Rationals vs Byzantines in consensus-based blockchains. Working paper, 2019. https://arxiv.org/abs/1902.07895. 169

N. Arnosti and M. Weinberg. Bitcoin: A natural oligopoly. Working paper, 2018. https://arxiv.org/pdf/1811.08572.pdf. 162

S. Athey, I. Parashkevov, V. Sarukkai and J. Xia. Bitcoin pricing, adoption, and usage: Theory and evidence. Stanford University Graduate School of Business Research Paper No. 16-42, 2016. 157

C. Badertscher, J. Garay, U. Maurer, D. Tschudi, and V. Zikas. But why does it work? A rational protocol design treatment of Bitcoin. *EUROCRYPT (2)*, pages 34–65. Springer, 2018. 156

C. Barrera and S. Hurder. Blockchain upgrade as a coordination game. Working paper, 2018. http://papers.ssm.com/sol3/papers.cfm?abstract_id=31922008. 160

J. Beccuti and C. Jaag. The Bitcoin mining game: On the optimality of honesty in proof-of-work consensus mechanism. Swiss Economics Working Paper 0060, 2017. 161

B. Biais, C. Bisière, M. Bouvard, and C. Casamatta. The blockchain folk theorem. *The Review of Financial Studies* 32(5), 1662–1715, 2019a. 157, 158, 159, 162

B. Biais, C. Bisière, M. Bouvard, and C. Casamatta. Blockchains, coordination and forks. *AEA Papers and Proceedings* 109, 88–92, 2019b. 159, 160

B. Biais, C. Bisière, M. Bouvard, C. Casamatta, and A. Menkveld. Equilibrium Bitcoin pricing. TSE working paper 18–73. 2019c. 157

J. Bonneau, E. W. Felten, S. Goldfeder, J. A. Kroll, and A. Narayanan. Why buy when you can rent? Bribery attacks on Bitcoin-style consensus. *3rd Workshop on Bitcoin and Blockchain Research (BITCOIN 2016), Barbados*, pages 19–26. Springer Verlag, 2016. 159

J. Brown-Cohen, A. Narayanan, C. A. Psomas, and S. M. A. Weinberg. Formal barriers to longest-chain proof-of-stake protocols. Cornell University working paper, 2018. http://arxiv.org/abs/1809.06528. 169

M. Carlsten, H. Kalodner, S. M. Weinberg, and A. Narayanan. On the instability of Bitcoin without the block reward. In *Proceedings of the 2016 ACM SIGSAC Conference on Computer and Communications Security*, pages 154–167. ACM, 2016. 158, 159

J. Chiu and T. Koeppl. The economics of cryptocurrencies—Bitcoin and beyond. 2017. Available at SSRN: http://dx.doi.org/10.2139/ssrn.3048124. 166

J. Chiu and T. V. Koeppl. Blockchain-based settlement for asset trading. *The Review of Financial Studies* 32(5), 1716–1753, 2019. 166, 167, 168

L. W. Cong, Z. He, and J. Li. Decentralized mining in centralized pools. *The Review of Financial Studies* 34(3), 1191–1235, 2021a. 164, 165

L. W. Cong, Y. Li, and N. Wang. Tokenomics: Dynamic adoption and valuation. *The Review of Financial Studies* 34(3), 1105–1155, 2021b. 157

N. Dimitri. Bitcoin mining as a contest. *Ledger* 2, 2017. http://doi.org/10.5195/ledger.2017.96. 162

D. Easley, M. O'Hara, and S. Basu. From mining to markets: The evolution of Bitcoin transaction fees. *Journal of Financial Economics* 134, 91–109, 2019. 166, 168

I. Eyal. The miner's dilemma. In *2015 IEEE Symposium on Security and Privacy (SP)*, pages 89–103. IEEE, 2015. 165, 166

I. Eyal and E. G. Sirer. Majority is not enough: Bitcoin mining is vulnerable. In N. Christin, and R. Safavi-Naini (editors), *Financial Cryptography and Data Security*, FC 2014, volume 8437 of *Lecture Notes in Computer Science*, pages 436–454. Springer, Berlin, Heidelberg, 2014. 160, 161, 165

G. Fanti, L. Kogan, and P. Viswanath. Economics of proof-of-stake payment systems. Working paper, 2019. https://pramodv.ece.illinois.edu/pubs/GKV.pdf. 157, 168

J. Fernández-Villaverde and D. Sanchez. On the economics of digital currencies. Federal Reserve Bank of Philadelphia working paper, 2018. https://papers.ssm.com/sol3/papers.cfm?abstract_id=3117347. 157

D. Ferreira, J. Li, and R. Nicolowa. Corporate capture of blockchain governance. Working paper, 2019. https://papers.ssm.com/sol3/papers.cfm?abstract_id=3320437. 165

B. Fisch, R. Pass, and A. Shelat. Socially optimal mining pools. *International Conference on Web and Internet Economics*, pages 205–218. Springer, 2017. 164

N. Gandal and H. Halaburda. Can we predict the winner in a market with network effects? Competition in cryptocurrency market. *Games* 7, 1–21, 2016. 157

R. Garratt and N. Wallace. Bitcoin 1, Bitcoin 2 . . . : An experiment with privately issued outside monies. *Economic Inquiry* 56 (3), 1887–1897, 2018. 157

C. Grunspan and R. Pérez-Marco. On profitability of selfish mining. Working paper, 2019. https://arxivv .org/abs/1805.08281. 161

H. Halaburda and G. Haeringer. Bitcoin and blockchain: What we know and what questions are still open. NYU Stern School of Business, *Baruch College Zicklin School of Business Research Paper* No. 2018-10-02, 2019. Available at SSRN: https://ssrn.com/abstract=3274331. 156

N. Houy. The Bitcoin mining game. *Ledger* 1, 2016. https://doi.org/10.5195/ledger.2016.13. 159

G. Huberman, J. Leshno, and C. Moallemi. An economic analysis of the Bitcoin payment system. Working paper, 2018. https://econ.hkbu.edu.hk/eng/Doc/Bitcoin_Payment_System.pdf. 166, 167, 168

B. Johnson, A. Laszka, J. Grossklags, M. Vasek, and T. Moore. Game-theoretic analysis of DDoS attacks against Bitcoin mining pools. In *International Conference on Financial Cryptography and Data Security*, pages 72–86. Springer, 2014. 165, 166

A. Kiayias, E. Koutsoupias, M. Kyropoulou, and Y. Tselekounis. Blockchain mining games. In *Proceedings of the 2016 ACM Conference on Economics and Computation*, pages 365–382. ACM, 2016. 157, 158, 159

J. A. Kroll, I. C. Davey, and E. W. Felten. The economics of Bitcoin mining, or Bitcoin in the presence of adversaries. In *Proceedings of WEIS, June 11–12, 2013, Washington, DC*, pages 1–21. 2013. 157, 158

Z. Liu, N. C. Luong, W. Wang, D. Niyato, P. Wang, Y-C. Liang, and D. I. Kim. A survey on applications of game theory in blockchain, 2019. https://arxiv.org/pdf/1902.10865.pdf. 156

J. Ma, J. Gans, and R. Tourky. Market structure in Bitcoin mining. NBER working paper 24242, 2018. https://www.nber.org/papers/w24242. 162

M. Manshaei, M. Jadliwala, A. Maiti, and M. Fooladgar. A game-theoretic analysis of shard-based permissionless blockchains. *IEEE Access* 6:78100–78112, 2018. 169

S. Nakamoto. Bitcoin: A peer-to-peer electronic cash system, 2008. https://bitcoin.org/bitcoin.pdf. 155, 157

E. Pagnotta. Bitcoin as decentralized money: Prices, mining, and network security. Working paper, 2018. https://www.jbs.cam.ac.uk/wp-content/uploads/2020/08/paper-pagnotta-bitcoinasdecentralised money.pdf. 157, 163

J. Prat and B. Walter. An equilibrium model of the market for Bitcoin mining. Working paper, 2018. https://papers.ssm.com/sol3/papers.cfm?abstract_id=3143410. 162, 163

M. Rosenfeld. Analysis of Bitcoin pooled mining reward systems. Working paper, 2011. https://arxiv.org/ abs/1112.4980. 164

F. Saleh. Volatility and welfare in a crypto economy. Working paper, 2018. https://papers.ssm.com/sol3/ papers.cfm?abstract_id=3235467. 157

F. Saleh. Blockchain without waste: Proof-of-stake. *The Review of Financial Studies* 34(3), 1156–1190, 2021. 168, 169

A. Sapirshtein, Y. Sompolinsky, and A. Zohar. Optimal selfish mining strategies in Bitcoin. In *International Conference on Financial Cryptography and Data Security*, pages 515–532. Springer, 2016. 161

O. Schrijvers, J. Bonneau, D. Boneh, and T. Roughgarden. Incentive compatibility of Bitcoin mining pool reward functions. In J. Grossklags and B. Preneel, editors, *Financial Cryptography and Data Security, FC 2016*, volume 9603 of *Lecture Notes in Computer Science*, pages 477–498. Springer, Berlin, Heidelberg, 2017. 164

L. Schilling and H. Uhlig. Some simple Bitcoin economics. *Journal of Monetary Economics* 106, pages 16–26, 2019. 157

J. Teutsch, S. Jain, and P. Saxena. When cryptocurrencies mine their own business. In J. Grossklags and B. Preneel, editors, *Financial Cryptography and Data Security, FC 2016*, volume 9603 of *Lecture Notes in Computer Science*, pages 499–514. Springer, Berlin, Heidelberg, 2016. 159

I. Tsabary and I. Eyal. The gap game. In *Proceedings of the 2018 ACM SIGSAC Conference on Computer and Communications Security*, pages 713–728, 2018. 167

Y. Zhu and S. Hendry. A framework for analyzing monetary policy in an economy with e-money. Bank of Canada Staff working paper, 2019. https://www.bankofcanada.ca/wp-content/uploads/2019/01/swp2019-1.pdf. 157

AUTHORS' BIOGRAPHIES

Bruno Biais holds a PhD in finance from HEC, and has received both the Paris Bourse dissertation award and the CNRS bronze medal. He taught at Toulouse, Carnegie Mellon, Oxford, and LSE; was Research Director at CNRS; and is now Professor at HEC. His research on finance, contract theory, political economy, experimental economics, and blockchains has been published in *Econometrica, JPE, AER, Review of Economic Studies, Journal of Finance, RFS*, and *JFE*. He was editor of the *Review of Economic Studies* and of the *Journal of Finance*, and he is currently departmental editor for finance at *Management Science*. He is a fellow of the Econometric Society, the Society for the Advancement of Economic Theory, and the Finance Theory group. He has been scientific adviser to the NYSE, Euronext, the European Central Bank, and the Bank of England. His work on trading and post trading received a senior ERC grant, and his current work on welfare incentives dynamics and equilibrium is being funded by a senior ERC grant.

Christophe Bisière is a professor of finance at Toulouse Capitole University (TSE and TSM). He holds a research master's degree in computer science, and a research master's degree in finance. He received his PhD in economics from Aix-Marseille University in 1994. He was an economic fellow of the U.S. Securities and Exchange Commission in Washington, DC, working on U.S. markets regulation issues. His research on finance, experimental economics, and blockchain has been published in international academic journals such as *Computational Economics, European*

Financial Management, Management Science, and the *Review of Financial Studies*. He was co-organizer of Tokenomics 2020, the 2nd International Conference on Blockchain Economics, Security and Protocols.

Matthieu Bouvard is Professor of Finance at Toulouse Capitole University (TSE and TSM). He obtained his PhD from the University of Toulouse in 2009 and was on the faculty at McGill University from 2009 to 2019. His research focuses on financial intermediation and the impact of technological innovation in finance (Fintech). His work has been published in academic finance journals such as the *Journal of Finance*, the *Journal of Financial Economics*, and the *Review of Financial Studies*. He has received funding from the National Research Agency in France (ANR), and from several funding agencies in Canada (FRQSC, IFSID, SSHRC). He was co-organizer of Tokenomics 2020, the 2nd International Conference on Blockchain Economics, Security and Protocols.

Catherine Casamatta is Professor of Finance at Toulouse Capitole University (TSE and TSM). She holds a master's degree from ESSEC and a PhD in finance from Toulouse Capitole University. Her research on corporate finance and governance, venture capital, and digital finance has been published in international academic journals such as the *American Economic Journal: Microeconomics*, the *Journal of Finance*, the *Journal of Financial Intermediation*, the *Journal of Money, Credit and Banking*, the *Review of Finance*, and the *Review of Financial Studies*. She is now the Toulouse Capitole University Vice President for Finance. She was co-organizer of Tokenomics 2020, the 2nd International Conference on Blockchain Economics, Security and Protocols.

CHAPTER 7

Bankruptcy Solutions as Reward Functions in Mining Pools

Marianna Belotti, *Cedric, Cnam, BDTD60, Caisse des Dépôts, Paris, France*

Stefano Moretti, *LAMSADE, CNRS, Université Paris-Dauphine, Université PSL, Paris, France*

7.1   INTRODUCTION

In the Bitcoin blockchain, transactions are collected in blocks, validated, and published on the ledger. Nakamoto [2008] proposed a system that validates blocks through a proof-of-work mechanism and chains them one to another. The proof-of-work mechanism requires finding an input that, when hashed with a *hashing function*, meets the difficulty target. More precisely, the goal for the block's validators (*miners*) is to find a numerical value (*nonce*) that, added to an input data string and hashed, gives an output that is lower than the target. A miner who finds a nonce meeting the difficulty target (*full solution*) spreads it across the network.

Miners compete to be the first to find a full solution, in order to publish the corresponding block and gain the Bitcoin reward. Mining is a competition among miners to solve complex cryptographic puzzles as fast as possible. The difficulty D of the puzzle ensures that it takes approximately 10 minutes to find a full solution. This difficulty value is adjusted periodically in order to meet the established validation rate and at the time of writing, $D \approx 20, 61 \cdot 10^{12}$.

7.1.1   MINING AND POOLING

Mining is a procedure through which miners can gain a substantial amount of money. As more people started mining in 2010, the difficulty of the proof-of-work procedure increased exponentially. Mining is now a professional activity requiring a substantial financial investment in competitive devices. Nowadays, due to the high difficulty values, mining with traditional devices takes a very long time before gaining a reward. On average, a miner who works alone with a personal laptop finds a full solution after billions of years. Small miners survive in this new industry by joining mining pools.

A mining pool is a cooperative approach where multiple miners share their efforts in order to validate blocks and gain rewards. Within a mining pool, several agents aggregate their computational power to solve crypto-puzzles. Then, the reward is split among them according to their contribution. Small miners, instead of waiting for years before gaining a reward, may in this way get a fraction of a reward but on a regular basis.

During the mining race, miners are connected to the pool's server(s) and their equipment is connected and synchronized with other miners' hardware. Miners participating in a pool share efforts and rewards. Miners' rewards are based on their contributions to finding a full solution. However, how do mining pools measure the work that members perform? The idea is to apply the proof-of-work algorithm to a "simpler" crypto-problem. More precisely, pools assign miners a problem to solve, setting a lower difficulty. Miners have to find the nonce of a simpler problem obtained by raising the threshold value. The solutions of this simpler crypto-puzzle are considered "near to valid" solutions and called *shares*. For the Bitcoin blockchain, adopting the SHA-256 hashing function, every hash value is a full solution with probability $\frac{1}{2^{32}D}$, and each hash has a probability of $\frac{1}{2^{32}}$ to be a share [Rosenfeld 2011].

Miners are rewarded according to the number of shares that they provide. Whenever a share is also a full solution, a block is validated and the pool gains a reward that is split among pool participants according to the number of shares that they have reported. A share is a full solution with probability $\frac{1}{D}$.

Mining pools are managed by a pool manager that establishes the way in which miners can be rewarded. More precisely, each pool adopts its own reward system. There are several reward approaches that can be more or less attractive to miners. Pool managers usually deduct a percentage fee of the reward before splitting it among miners.

7.1.2   GAME THEORY IN THE BITCOIN BLOCKCHAIN

Game theory is the branch of mathematics that studies the optimal decision-making processes in the presence of multiple decision-makers (i.e., *players*). The latter are considered to be *rational* agents, meaning that their aim is to maximize their own utility function, which does not always coincide with an egoistic choice. A *game* consists of an interactive decision-making process involving a set of players and a set of possible strategies accompanied by the corresponding outcomes. Several situations may be modeled as games leading to a game classification. Two main classes of games are: (i) cooperative games and (ii) non-cooperative games. While the latter class is devoted to predict players' individual strategies, the former explores games in which there exists the possibility for players to make binding agreements. The goal of cooperative game theory is to predict the formation of players' coalitions, and to study their stability.

As surveyed in Liu et al. [2019], game theory has been applied to the analysis of various issues related to Bitcoin and, more generally, to the blockchain. The vast majority of these applications are based on non-cooperative games, dealing with conflict situations where players cannot cooperate. Despite the fact that the problem of how to analyze the behavior of miners in mining pools is quite present in the Bitcoin literature [Conti et al. 2018], cooperative games and the theory of coalition formation have received much less attention so far. Due to the fact that mining pools are nothing but coalitions of rational players, we focus on analyzing the miners' behaviors when forming pools with cooperative game theory.

Mining pool attacks

An attack to a mining pool refers to any miner's behavior that differs from the default practice (the honest one) and that jeopardizes the collective welfare of the pool. Rosenfeld [2011] provided an overview of possible malicious behavior toward pools, which varies according to a given pool's reward mechanism. Miners may attack their pool at the time of reporting their proof of work. More precisely, they can (i) delay in reporting their shares (i.e., *block withholding*) and/or (ii) report shares elsewhere (i.e., *pool-hopping*) and/or (iii) report shares by appearing as different reporting entities (i.e., *sybil attack*).

The first of these attacks, block withholding, occurs when a miner intentionally delays in reporting shares and full solutions to a mining pool. This can result in delaying a block validation and the pool's consequent possession of the reward. In some cases it may be profitable for attackers to be rewarded after the submission of additional shares. Pool-hopping, the second kind of attack, occurs when miners "hop" from one pool to another according to pools' attractiveness. Working for different pools at different times could be highly profitable. The last of these, the sybil attack, is an attempt to fill the network with nodes controlled by a single person. In this case, an attacker will appear in the network as several clients and/or several miners that are less powerful.

The problem for the pool manager is now to establish how to redistribute the rewards among pool participants in order to prevent such malicious behaviors. In other words, the pool manager must choose an "appropriate" reward mechanism preventing (possibly all) different types of attack.

As far as we know, the only paper using cooperative games to study under which conditions miners wish to form pools is Lewenberg et al. [2015], where the authors show that agents are interested in hopping between pools as a consequence of high transaction loads in the network processes. These authors claim that pool-hopping is not preventable and therefore that mining pools cannot be considered as stable coalitions; this is due to the fact that there is no reward system that can prevent this malicious behavior. The results provided in the paper strongly rely on the notion of *core* stability for coalitions (pools) within cooperative games. However, existing pools' reward systems are actually very basic [Schrijvers et al. 2016], so the assumption that any core allocation can be obtained as the outcome of a negotiation process within a pool is quite a strong hypothesis to have been made in Lewenberg et al. [2015].

In Schrijvers et al. [2016], the authors make use of non-cooperative game theory to propose a reward mechanism that prevents miners from withholding full solutions. In this book chapter, as well as in Belotti et al. [2020], we show that a proposed reward function can be reinterpreted as a solution of a bankruptcy game in a situation where the reward gained by the pool is not sufficient for paying out all the miners that have contributed (i.e., that have reported at least one share). Section 7.2 presents the model of Schrijvers et al. [2016]. In Section 7.3 we introduce some basic notions and definitions from the literature on game theory and bankruptcy situations. In Section 7.4 we extend (over the long run) the definition of the reward function introduced in Schrijvers et al. [2016] to some classical bankruptcy solutions, and we analyze their main properties in the Bitcoin context.

7.2   AN INCENTIVE-COMPATIBLE REWARD FUNCTION

The model considered is a simple one where a fixed number of miners n work actively for a particular pool. This model is based on a division into *rounds*. During a round, miners participate in the mining race and report their shares and full solutions to the pool manager. At the time when a full solution is submitted, the pool manager communicates it to the Bitcoin network and receives the block reward B (to simplify the exposition, $B = 1$). Then, the pool manager redistributes the block reward B among the n miners according to a pre-defined reward function. The round is then concluded and a new mining race starts.

In particular, miners' strategic behavior consists of deciding when to report a full solution to the pool manager. To represent such a situation, the authors introduce the notion of a *history transcript* as a vector $\mathbf{s} = (s_1, s_2, \ldots, s_n) \in \mathbb{N}^N$ containing the number of shares reported by each miner in the round. Letting $\mathbf{S} = \sum_{i \in N} s_i$ be the total number of reported shares, the reward function $R : \mathbb{N}^N \to [0, 1]^n$, according to Schrijvers et al. [2016], is a function assigning to each history transcript $\mathbf{S}$ an allocation of the reward $(R_1(\mathbf{S}), \ldots, R_n(\mathbf{S}))$, where R_i denotes the fraction of reward gained by the single miner $i \in N$ and $\sum_{i \in N} R_i = B = 1$. Notice that the reward function considered in Schrijvers et al. [2016] is independent from the order in which shares are submitted to the pool manager.

As already mentioned, in this simple model miners have only two possible strategies: (i) immediately reporting a full solution, or (ii) delaying the report. The first option is considered to be the *default mining strategy*, while the second has to be considered as an *attack* for the pool, as it can compromise its total reward.

In Schrijvers et al. [2016], a reward function R is said to be *incentive compatible* whenever every (rational) miner's best strategy (i.e., the strategy that produces the most favorable outcome for an agent) is to immediately report a share or a full solution; that is, the reward expected by adopting the strategy to report a share or a full solution immediately is larger than the reward expected by adopting the strategy to delay their report. Formally, the incentive-compatible map R^w proposed in Schrijvers et al. [2016] is defined as follows:

$$R_i^w(\mathbf{s}) = \begin{cases} \frac{s_i}{D} + e_i^w \left(1 - \frac{S}{D}\right) & \text{if } S < D \\ \frac{s_i}{S} & \text{if } S \geq D, \end{cases} \qquad \forall i \in N \qquad (7.1)$$

where $\mathbf{e}^w = (e_1^w, \ldots, e_n^w) \in \{0, 1\}^N$ is a vector such that $e_i^w = 0$ for each $i \in N \setminus \{w\}$ and $e_w^w = 1$. So, in case $S \geq D$, the reward function is proportional to the submitted shares, whereas in case $S < D$ each miner receives a fixed reward per share equal to $\frac{1}{D}$ and the discoverer w of the full solution receives, in addition, all the remaining amount $1 - \frac{S}{D}$. So, in any case, $\sum_{i \in N} R_i^w(\mathbf{s}) = B = 1$.

Roughly speaking, the reward function R^w is the combination of two distinct allocation methods. In the *short run*—that is, when the total amount of reported shares is smaller than the difficulty D of the original problem—the reward function allocates a fixed amount per share to all agents equal to $\frac{1}{D}$, but the agent w who finds a solution is rewarded with an extra prize. Instead, in the *long run*—i.e., when the total amount of reported shares exceeds the difficulty of the problem—the reward function allocates the reward proportionally to the individual shares.

In the following sections, we make a parallel between the behavior of the reward function R^w in the long run and the classical proportional allocation rule for *bankruptcy situations*, and we will extend the analysis of such a reward function to other well-known solutions for bankruptcy situations.

7.3   COOPERATIVE GAME THEORY AND BANKRUPTCY SITUATIONS

The theory on cooperative games provides mathematical models that can describe several realistic situations where working together allows people to achieve their goals more easily than trying to pursue them alone.

Let us denote by N a finite set of players and by S any subset of N (i.e., a coalition); cooperative games—also called transferable utility (TU) games—are defined by the couple (N, v) where $v : 2^N \to \mathbb{R}$, $v(\emptyset) = 0$ is the *characteristic function* of the game. This function associates a real value to all the possible coalitions. More precisely, $v(S)$ represents the value that each coalition S can get for itself, once formed. A solution of a cooperative game (N, v) is a vector of the type $(x_1, \ldots, x_n)$, where x_i represents the payoff of player i. Solutions of cooperative games are asked (i) to assign to each player not less than what they can get on their own and (ii) to distribute all the available amount among the players. Solutions of a game satisfying the two properties are called *imputations* and are denoted as $\mathcal{I}(v)$.

Cooperative game theory is interested in identifying those imputations that make the *grand coalition* (i.e., the coalition formed by all the players in the game) stable under deviation. More precisely, if any coalition has no incentive to leave the grand coalition and form a sub-coalition, then the imputation is *core stable*.

Definition 7.1 (The core)   Let $v : 2^N \to \mathbb{R}$ be a TU game. The core of the game, denoted by $\mathcal{C}(v)$, is the set:

$$\mathcal{C}(v) = \left\{ x \in \mathbb{R}^n : \sum_{i \in N} x_i = v(N) \wedge \sum_{i \in S} x_i \geq v(S), \quad \forall S \subset N \right\}.$$

The core $\mathcal{C}(v)$ is a subset of the imputation set $\mathcal{I}(v)$ containing the imputations that are not rejected by any coalition. Each coalition with associated payoffs in the core receives at least what it deserves; therefore no subset of players has any incentive to leave the grand coalition. In Definition 7.1 it is the following inequality,

$$\sum_{i \in S} x_i \geq v(S), \quad \forall S \subset N,$$

that represents a stability concept guaranteeing that agents with offered payoffs in the core stay in the grand coalition and cooperate together.

7.3.1   BANKRUPTCY SITUATIONS

A bankruptcy situation arises whenever there are some agents claiming a certain amount of a divisible estate, and the sum of the claims is larger than the estate. Let $N = \{1, \ldots, n\}$ be a set of agents. Formally, a *bankruptcy situation* on the set N consists of a pair $(\mathbf{c}, E) \in \mathbb{R}^N \times \mathbb{R}$ with $c_i \geq 0$ for all $i \in N$ and $0 < E < \sum_{i \in N} c_i = C$. The vector $\mathbf{c}$ represents agents' demands (each agent $i \in N$ claims a quantity c_i) and E is the estate that has to be divided among agents (and it is not enough to satisfy the total demand C).

Denote by $\mathbb{B}^N$ the class of all bankruptcy problems $(\mathbf{c}, E) \in \mathbb{R}^N \times \mathbb{R}$ with $0 < E < \sum_{i \in N} c_i$ and N as a set of agents. A *solution* (also called an *allocation rule*) for bankruptcy situations on N consists of a map $f : \mathbb{B}^N \to \mathbb{R}^N$ assigning to each bankruptcy situation in $\mathbb{B}^N$ an *allocation vector* in $\mathbb{R}^N$, which specifies the amount $f_i(\mathbf{c}, E) \in \mathbb{R}$ of estates E that each player $i \in N$ receives in situation $(\mathbf{c}, E)$. A solution must satisfy a minimal set of "natural" requirements:

Individual rationality. $f_i(\mathbf{c}, E) \geq 0 \; \forall i \in N$, saying that every agent must receive a non-negative amount of the estate E;

Demands boundedness. $f_i(\mathbf{c}, E) \leq c_i \ \forall i \in N$, stating that no agent receives strictly more than what each claims;

Efficiency. $\sum_{i \in N} f_i(\mathbf{c}, E) = E$, requiring that the entire estate must be allocated among the claiming agents.

We now introduce three well-studied solutions for bankruptcy situations (see, for instance, Moulin [2000], Herrero and Villar [2001], Thomson [2003], Curiel et al. [1987], O'Neill [1982]). The first one is the *proportional rule*.

Definition 7.2 (Proportional rule (P))   For each bankruptcy situation $(\mathbf{c}, E) \in \mathbb{B}^N$, the proportional rule yields the allocation vector $P(\mathbf{c}, E) = \pi \mathbf{c}$, where π is such that $\sum_{i \in N} \pi c_i = E$.

Example 7.3   Consider a bankruptcy problem with three claimants $N = \{1, 2, 3\}$ and such that $(\mathbf{c}, E) = ((4, 5, 3), 10)$. First, compute the parameter $\pi = \frac{E}{C} = \frac{10}{12} = \frac{5}{6}$, then get the proportional allocation by multiplying π by the vector of demands $\mathbf{c}$:

$$P(\mathbf{c}, E) = \frac{5}{6}(4, 5, 3).$$

The second solution that we consider from the literature is the *constrained equal awards rule*. This allocating rule ignores differences among claimants and, thanks to the boundedness constraint, rules out cases where agents receive more than what they claim.

Definition 7.4 (Constrained equal awards rule (CEA))   For each bankruptcy situation $(\mathbf{c}, E) \in \mathbb{B}^N$, the constrained equal awards rule is defined as $CEA_i(\mathbf{c}, E) = \min\{c_i, \lambda\}$, where the parameter λ is such that $\sum_{i \in N} \min\{c_i, \lambda\} = E$.

The CEA rule consists of giving to every agent the same amount until the agent's demand is not completely satisfied and the estate is not finished.

Example 7.5   Consider the bankruptcy situation given in Example 7.3. The first step assigns to all players $x = (3, 3, 3)$. The game to solve is now: $(\mathbf{c}', E') = ((1, 2, 0), 1)$. Since the third agent is satisfied, E' has to be allocated between the first two players. Here $\lambda = \frac{7}{2}$; hence,

$$CEA(\mathbf{c}, E) = \left(\frac{7}{2}, \frac{7}{2}, 3\right).$$

The last division rule we consider, the *constrained equal losses rule*, is an alternative to the CEA rule and focuses on the losses claimants incur. More precisely, instead of equating awards, this rule equates losses.

Definition 7.6 (Constrained equal losses rule (CEL))   For each bankruptcy situation $(\mathbf{c}, E) \in \mathbb{B}^N$, the constrained equal losses rule is defined as $CEL_i(\mathbf{c}, E) = \max\{c_i - \lambda', 0\}$, where the parameter λ' is such that $\sum_{i \in N} \max\{c_i - \lambda', 0\} = E$.

Example 7.7   Consider again the bankruptcy situation given in Example 7.3. At first, a loss of 1 unit is assigned to each player: $x = (3, 4, 2)$. Then, it is easy to check that $\lambda' = \frac{1}{3}$ and therefore we have $(4, 5, 3) - \frac{1}{3} = (\frac{10}{3}, \frac{13}{3}, \frac{7}{3})$. So,

$$CEL(\mathbf{c}, E) = \left(\frac{10}{3}, \frac{13}{3}, \frac{7}{3} \right).$$

7.3.2   BANKRUPTCY GAMES: GAME-THEORETIC DIVISION RULES

Bankruptcy problems are not necessarily bankruptcy games; they can be solved without using game theory—that is, by applying a natural allocating rule such as the proportional one. Moreover, division rules are not necessarily game-theoretic division rules. However, these types of problems where an estate is divided among claimants suggest a cooperative behavior. Therefore, it is natural to try to model bankruptcy problems as a cooperative game and analyze the relative solutions. The first step is defining a characteristic function.

It is clear that the grand coalition must form and that $v(N) = E$, but it is not clear which value should be associated to single players and to intermediate coalitions. From the literature there exists a "pessimistic" approach to evaluate a coalition; the pessimistic characteristic function assigns to a coalition a value which is what S gets once the other $N \setminus S$ players have already taken what they claim. The coalition gets the part of the estate that is left after all other players are fully rewarded:

$$v_p(S) = \max \left(0, E - \sum_{i \in N \setminus S} c_i \right).$$

Bankruptcy problems can be represented as cooperative games with pessimistic characteristic functions. At this point, it is natural to ask: *what about division rules associated with this type of games?* The authors in Curiel et al. [1987] proposed a necessary and sufficient condition for a division rule to be a game-theoretic division rule:

Theorem 7.8 (Curiel 1987)   *An allocating rule $f(\mathbf{c}, E)$ for bankruptcy problems is a game-theoretic division rule if and only if $f(\mathbf{c}, E) = f(\mathbf{c}^T, E)$, $c_i^T = \min\{c_i, E\}$ for all $i \in N$.*

This condition derives from the fact that cooperative games corresponding to the bankruptcy problems $(\mathbf{c}, E)$ and $(\mathbf{c}^T, E)$ are the same. According to this theorem, neither the constrained equal losses nor the proportional rules are game-theoretic rules. For the CEL rule, it is not so immediate to conclude this as for the proportional one. Let us show this fact with a simple example:

Example 7.9   Let $(\mathbf{c}, E) = ((2, 11), 10)$ and so $(\mathbf{c}^T, E) = ((2, 10), 10)$. We have two different outcomes for the problems:

$$CEL(\mathbf{c}, E) = (0.5, 9.5), \qquad CEL(\mathbf{c}^T, E) = (1, 9).$$

In the case of the constrained equal awards rule, however, we can immediately see that it is a game theoretic division rule by simply using the rule's definition: $CEA_i(\mathbf{c}^T, E) = \min\{c_i, E, \lambda\} = \min\{c_i, \lambda\} = CEA_i(\mathbf{c}, E)$ since $\lambda \in [0, c_n]: \lambda \leq E$.

Another interesting consideration regarding the solutions of bankruptcy problems is the fact that the core of the associated pessimistic games coincides with the set of admissible solutions.

Proposition 7.10   *Given a bankruptcy problem* $(\mathbf{c}, E) \in \mathbb{B}^N$ *and its associated pessimistic cooperative game* $v_{p(\mathcal{C}, E)}$, *calling x the outcome of the problem/game, we have that:*

$$x \in \mathcal{C}(v_p) \Longleftrightarrow \begin{cases} \sum_{i \in N} x_i = E \\ 0 \leq x_i \leq c_i, \ \forall i \in N. \end{cases}$$

Proof. Concerning the *only if* part, the first condition is efficiency. For the second one we get $\forall i \in N, x_i \geq v_p(i) \geq 0$ and $E - x_i = \sum_{j \in N \setminus \{i\}} x_j \geq v_p(N \setminus \{i\}) \geq E - c_i$, which implies $x_i \leq c_i$. In order to prove the *if* part, let us notice that the efficiency condition is satisfied. Moreover, $\forall S \subseteq N$ we have two cases:

$$v_p(S) = 0 \leq \sum_{i \in S} x_i$$

$$v_p(S) = E - \sum_{i \in N \setminus S} c_i \leq E - \sum_{i \in N \setminus S} x_i \leq \sum_{i \in S} x_i.$$

$\square$

This theorem states that every game-theoretic division rule of bankruptcy problems provides allocations that are in the core of the associated cooperative games. Therefore, given a bankruptcy game $(\mathbf{c}, E)$, the constrained equal awards rule provides an outcome contained in the set $\mathcal{C}(v_{(\mathbf{c}, E)})$.

7.3.3   BANKRUPTCY RULES' PROPERTIES

Following the presentation in Herrero and Villar [2001], we now focus on some properties from the literature that are satisfied by the three previously introduced solutions, and that will be relevant for the discussion about Bitcoin systems in the next section. The first property consists of a basic equity requirement that players with identical claims are identical for the allocation rule. More precisely, agents with the same demands are treated identically in the sense that they receive the same fraction of the estate.

Property 7.11 (Equal treatment of equals)   For all $(\mathbf{c}, E) \in \mathbb{B}^N$ and all $i, j \in N$ we have that $c_i = c_j$ implies

$$f_i(\mathbf{c}, E) = f_j(\mathbf{c}, E).$$

The second property requires that an allocation rule is invariant to scale change, so it does not depend on the units in which problems' demands and estate are expressed.

Property 7.12 (Scale invariance)   For all $(\mathbf{c}, E) \in \mathbb{B}^N$ and all $\gamma > 0$ we have:

$$f(\gamma \mathbf{c}, \gamma E) = \gamma f(\mathbf{c}, E).$$

The next property refers to a situation where the population of agents may change. This property considers the case in which, after providing an allocation vector according to a solution applied to the original situation $(\mathbf{c}, E) \in \mathbb{B}^N$, a group of agents $S \subset N$ forms and re-applies the solution to the reduced situation $(\mathbf{c}_S, \sum_{i \in S} f_i(\mathbf{c}, E)) \in \mathbb{B}^S$, where $\mathbf{c}_S = (c_i)_{i \in S}$ and the new estate is $\sum_{i \in S} f_i(\mathbf{c}, E)$. This property states that a solution applied to any reduced problem should provide to agents in S the same output as in the original situation.

Property 7.13 (Consistency)   For all $S \subset N$, all $(\mathbf{c}, E, N) \in \mathbb{B}$, and all $i \in S$ we have:

$$f_i(\mathbf{c}, E, N) = f_i(\mathbf{c}_S, \sum_{i \in S} f_i(\mathbf{c}, E, N), S).$$

The following two properties characterize the allocation priorities of an allocation rule. Exemption establishes that when claims are smaller than the equal division, the rule has to first satisfy them and therefore cut larger claims.

Property 7.14 (Exemption)   For all $(\mathbf{c}, E) \in \mathbb{B}^N$, if $c_i \leq E/n$, then $f_i(\mathbf{c}, E) = c_i$.

Exclusion gives an opposite message, since it excludes from the problem those agents with very low claims.

Property 7.15 (Exclusion)   For all $(\mathbf{c}, E) \in \mathbb{B}^N$, if $c_i \leq L/n$ where $L = \sum_{i=1}^{n} c_i - E$, then $f_i(\mathbf{c}, E) = 0$.

The next property deals with the possibility that a group of agents may aggregate their demands and appear as a single claimant, or that a single agent may split her demand to represent several claimants. If these behaviors are not beneficial for the agents, we say that a solution satisfies no advantageous merging or splitting.

Property 7.16 (No advantageous merging or splitting)   Let $(\mathbf{c}, E) \in \mathbb{B}^N$ and $(\mathbf{c}', E') \in \mathbb{B}^{N'}$ be such that $N' \subset N$, $E = E'$, and suppose there is an agent $i \in N'$ such that $c_i' = c_i + \sum_{j \in N \setminus N'} c_j$

and $c'_j = c_j$ for each $j \in N' \setminus \{i\}$; then

$$f_i(\mathbf{c}', E') = f_i(\mathbf{c}, E) + \sum_{j \in N \setminus N'} f_j(\mathbf{c}, E).$$

It is possible to split this property into two distinct properties by considering the two different behaviors of merging and splitting.

Property 7.17 (No advantageous merging)   Let $(\mathbf{c}, E) \in \mathbb{B}^N$ and $(\mathbf{c}', E') \in \mathbb{B}^{N'}$ be such that $N' \subset N$, $E = E'$, and suppose there is an agent $i \in N'$ such that $c'_i = c_i + \sum_{j \in N \setminus N'} c_j$ and $c'_j = c_j$ for each $j \in N' \setminus \{i\}$; then

$$f_i(\mathbf{c}', E') \leq f_i(\mathbf{c}, E) + \sum_{j \in N \setminus N'} f_j(\mathbf{c}, E).$$

Property 7.18 (No advantageous splitting)   Let $(\mathbf{c}, E) \in \mathbb{B}^N$ and $(\mathbf{c}', E') \in \mathbb{B}^{N'}$ be such that $N' \subset N$, $E = E'$, and suppose there is an agent $i \in N'$ such that $c'_i = c_i + \sum_{j \in N \setminus N'} c_j$ and $c'_j = c_j$ for each $j \in N' \setminus \{i\}$; then

$$f_i(\mathbf{c}', E') \geq f_i(\mathbf{c}, E) + \sum_{j \in N \setminus N'} f_j(\mathbf{c}, E).$$

According to Property 7.17 no group of agents has an incentive to aggregate their demands and to be treated as a single agent whose claim is the sum of the individual demands of their participants. Conversely, Property 7.18 says that no agent has an incentive to divide his/her demand to represent several claimants whose demands add up to her/his original claim. Solutions introduced in this section satisfy alternative sets of properties, as reported in Table 7.1. (For more details on these results see, for instance, Herrero and Villar [2001].)

TABLE 7.1: Three bankruptcy solutions and some of their properties

Properties	CEA	CEL	P
Equal treatment of equals	Yes	Yes	Yes
Scale invariance	Yes	Yes	Yes
Consistency	Yes	Yes	Yes
Exemption	Yes	No	No
Exclusion	No	Yes	No
No advantageous merging or splitting	No	No	Yes
No advantageous merging	Yes	No	Yes
No advantageous splitting	No	Yes	Yes

7.4   REWARD FUNCTIONS BASED ON BANKRUPTCY RULES

In this section we provide a game-theoretic representation for the incentive-compatible rule proposed in Schrijvers et al. [2016] for long runs, and we propose two alternative reward functions for mining pools based on the use of the CEA and CEL solutions, respectively, in the role of the proportional solution in equation (7.1) over the long run (i.e., $\mathbf{S} \geq D$).

7.4.1   GAME-THEORETIC REPRESENTATION FOR LONG RUNS

Given a bankruptcy problem (c, E) with n agents, its associated pessimistic cooperative game is $(v_{(c, E)}, N)$, where $N = \{1, \ldots, n\}$ and $v_{(c, E)}(S) = \max\big(E - \sum_{i \in N \setminus S} c_i, 0\big)$ for any coalition $S \subset N$. Therefore, we can associate to the bankruptcy problem $\mathcal{P} = \big\{\big(\frac{s_i}{D}, 1\big) : \mathbf{S} \geq D\big\}$ a cooperative game $v_{\mathcal{P}}$ where:

$$v_{\mathcal{P}}(S) = \max\left(1 - \frac{1}{D} \sum_{i \in N \setminus S} s_i, 0\right) \quad \forall S \subset N.$$

We know from Theorem 7.8 that the proportional rule is not a game-theoretic rule for problem $\mathcal{P}$. However, by simply adding a constraint on the claims $c_i \leq E$ (i.e., $s_i \leq D$ in the Bitcoin context[1]) the proportional rule is a game-theoretic division rule for the problem $\mathcal{P}' = \big\{\big(\frac{s_i}{D}, 1\big) : \mathbf{S} \geq D, s_i \leq D\big\}$. The proportional allocation is not only a solution of the associated cooperative game $v_{\mathcal{P}'}$ but it is also in the core of the game. Therefore, it is a stable solution of the bankruptcy game; that is, the proportional rule is not rejected by any coalition and there is no incentive for leaving a coalition. In the Bitcoin framework, stability of the allocating rule ensures that no miner leaves the pool to form sub-pools.

7.4.2   NEW REWARD FUNCTIONS

Since the incentive-compatible reward function R^w proposed in Schrijvers et al. [2016] can be modeled as a bankruptcy game for long runs,

$$R_i^w(\mathbf{s}) = \begin{cases} \frac{s_i}{D} + e_i^w \left(1 - \frac{\mathbf{S}}{D}\right) & \text{if } \mathbf{S} < D \\ \frac{s_i}{\mathbf{S}} & \text{if } \mathbf{S} \geq D, \end{cases} \quad \forall i \in N$$

we can construct two new reward functions by substituting for the proportional division the CEA and the CEL rules.

---

[1] The constraint on the claims is reasonable in the Bitcoin context. Having a situation with $s_i > D$ in a round means that all the computational power of the network, in that round, is concentrated in a single agent. Due to the decentralized nature of the network, this situation can be considered as extremely unlikely.

Definition 7.19   Let $w \in N$. The CEA-based reward function $\hat{R}^w$ is such that

$$\hat{R}_i^w(\mathbf{s}) = \begin{cases} \frac{s_i}{D} + e_i^w \left(1 - \frac{\mathbf{S}}{D}\right) & \text{if } \mathbf{S} < D \\ \min\left(\frac{s_i}{D}, \lambda\right) \text{ s.t. } \lambda : \sum_i \min\left(\frac{s_i}{D}, \lambda\right) = 1 & \text{if } \mathbf{S} \geq D. \end{cases} \quad \forall i \in N$$

Definition 7.20   Let $w \in N$. The CEL-based reward function $\bar{R}^w$ is such that

$$\bar{R}_i^w(\mathbf{s}) = \begin{cases} \frac{s_i}{D} + e_i^w \left(1 - \frac{\mathbf{S}}{D}\right) & \text{if } \mathbf{S} < D \\ \max\left(\frac{s_i}{D} - \lambda', 0\right) \text{ s.t. } \lambda' : \sum_i \max\left(\frac{s_i}{D} - \lambda', 0\right) = 1 & \text{if } \mathbf{S} \geq D. \end{cases} \quad \forall i \in N$$

Each of the proposed allocating rules solving the bankruptcy problem $\mathcal{P}'$ is also a game-theoretic division rule. Hence, all three reward functions are core-stable in long runs.

In Section 7.3.3 we have listed some common properties for solutions of bankruptcy situations. Some of these properties are satisfied by proportional, CEA, and CEL solutions (see Table 7.1), and therefore they are obviously also satisfied by R^w, $\hat{R}^w$, and $\bar{R}^w$, respectively, in the long run (i.e., $\mathbf{S} \geq D$). In the following, we discuss the interpretation of these properties in the Bitcoin framework.

Equal treatment of equals (7.11)

This property ensures impartiality: agents with the same number of shares have to be rewarded in the same way. In the Bitcoin framework, impartiality of the pool manager is mandatory. The pool operator knows neither each miner's computational power nor miners' identities. The goal of the reward system in use is to give an estimate of miners' hash power. Therefore, the number of reported shares s_i submitted by miner i is considered as an estimate of its computational power. Such protocols do not ask for a miner's identity in the submission procedure. In conclusion, equal treatment of equals establishes that miners with the same number of reported shares receive the same fraction of the total reward E.

Scale invariance (7.12)

This property ensures the independence of the reward function from the currency in which the problem's variables are defined. This property makes the reward function more robust to possible currency conversion during the mining activity, thus improving the stability of miners' interaction within the pool.

Consistency (7.13)

Consistency expresses the independence of a reward function with respect to population restrictions. In the Bitcoin framework, this property implies that no group of miners has an incentive to re-apply a solution in the reduced problem, and there is no advantage for miners to renegotiate the allocation in a sub-group once a reward function is accepted by the pool's participants.

No advantageous merging or splitting (7.16)

This property rules out the possibility for a group of miners to aggregate their shares in order to appear as a single miner, or conversely, the possibility for a single miner to split its share and appear as several miners. In the Bitcoin framework, this last behavior is identified as the sybil attack, presented in Section 7.1.2. These miners can create multiple accounts with different "usernames" and e-mail addresses and pretend to be different entities. The proportional rule is the only rule that satisfies the property of no advantageous merging or splitting. Therefore, it guarantees pools' robustness with respect to the sybil attack. However, in order to prevent this attack, it is enough to have an allocation rule satisfying only *no advantageous splitting*. As reported in Table 7.1, we have that the CEA-based reward function $\hat{R}^w$ satisfies *no advantageous merging (7.17)* for long runs, whereas the CEL-based reward function $\bar{R}^w$ satisfies *no advantageous splitting (7.18)* for long runs.

7.5   CONCLUSION

All the allocation rules R^w, $\hat{R}^w$, and $\bar{R}^w$ considered in this chapter satisfy, over the long run (i.e., if $\mathbf{S} \geq D$), the interesting properties of equal treatment of equals, scale invariance, and consistency.

However, only the rule R^w is incentive compatible, as proved in Schrijvers et al. [2016]. Indeed, neither the CEA nor the CEL rules are incentive compatible. Concerning CEA, Property 7.14 (exemption) tells us that every time an agent claims something less than E/n, its demand is fully satisfied. That is, a miner submitting $s_i : s_i \leq D/n$ shares is aware of being rewarded s_i/D even in the case of an intentional delay in reporting. For the CEL rule, due to Property 7.15 (exclusion), miners claiming more than D/n know that they will never receive their claim in full, but they cannot compare the expected rewards gained with the two different strategies (reporting and delaying their report).

All the rules can be interpreted as game-theoretic division rules of a bankruptcy problem for long runs. Furthermore, they are in the core of the cooperative game associated with the bankruptcy problem $\mathcal{P}'$. $\bar{R}^w$ and R^w satisfy no advantageous splitting; thus both of them are robust to sybil attacks. However, rule R^w is the only one that guarantees good behavior of agents within a given pool, since it is the only one that is incentive compatible.

Considering a multiple pool framework, miners can hop from one pool to another (pool-hopping attack) and can submit their share to several different pools.

In a situation where the reward function in use is $\hat{R}^w$, miners have an incentive to split their demands, since the CEA-based rule fails to satisfy no advantageous splitting and therefore there is an advantage for miners to split their shares. Moreover, if the claims (shares) become lower than $\frac{D}{n}$, miners are fully rewarded (Property 7.14), which results in a strong advantage to splitting big shares. More precisely, miners can split a big amount of shares into several small shares and report them all in one single pool, or in different pools, using the same reward system. We can thus conclude that the $\hat{R}^w$ rule does not prevent pool-hopping.

Conversely, if the reward function in use is $\bar{R}^w$ or R^w, miners seem to have no strict incentive for splitting their shares, as both solutions satisfy no advantageous splitting. However, using the R^w reward function, agents are indifferent between splitting or not (no advantageous merging or splitting property). Instead, using the $\bar{R}^w$ reward function, if the demands (shares) become lower than $\frac{L}{n} = \frac{S-1}{n}$ due to splitting, agents do not receive any positive reward (Property 7.15). So, the $\bar{R}^w$ reward function is the only one of the three rules that provides strong disincentives to malicious inter-pool behavior (by share splitting). The work in Belotti et al. [2020] presented a proper modification of $\bar{R}^w$ to preserve the incentive compatibility property.

BIBLIOGRAPHY

M. Belotti, S. Moretti, and P. Zappalà. Rewarding miners: Bankruptcy situations and pooling strategies. In *Multi-Agent Systems and Agreement Technologies*, pages 85–99. Springer, Cham. 2020. 178, 189

M. Conti, E. Sandeep Kumar, C. Lal, and S. Ruj. A survey on security and privacy issues of Bitcoin. *IEEE Communications Surveys & Tutorials* 20(4):3416–3452, 2018. 177

I. J. Curiel, M. Maschler, and S. H. Tijs. Bankruptcy games. *Zeitschrift für Operations Research* 31(5):A143–A159, 1987. 181, 182

C. Herrero and A. Villar. The Three Musketeers: Four classical solutions to bankruptcy problems. *Mathematical Social Sciences* 42(3):307–328, 2001. 181, 183, 185

Y. Lewenberg, Y. Bachrach, Y. Sompolinsky, A. Zohar, and J. S. Rosenschein. Bitcoin mining pools: A cooperative game theoretic analysis. In *Proceedings of the 2015 International Conference on Autonomous Agents and Multiagent Systems*, pages 919–927. Citeseer, 2015. 177, 178

Z. Liu, N. Luong, W. Wang, D. Niyato, P. Wang, Y. Liang, and D. Kim. A survey on applications of game theory in blockchain. In *arXiv preprint arXiv:1902.10865*, 2019. 177

H. Moulin. Priority rules and other asymmetric rationing methods. *Econometrica* 68(3):643–684, 2000. 181

S. Nakamoto. Bitcoin: A peer-to-peer electronic cash system, 2008. https://bitcoin.org/bitcoin.pdf. 175

B. O'Neill. A problem of rights arbitration from the Talmud. *Mathematical Social Sciences* 2(4):345–371, 1982. 181

M. Rosenfeld. Analysis of Bitcoin pooled mining reward systems. In *arXiv preprint arXiv:1112.4980*, 2011. 176, 177

O. Schrijvers, J. Bonneau, D. Boneh, and T. Roughgarden. Incentive compatibility of Bitcoin mining pool reward functions. In *International Conference on Financial Cryptography and Data Security*, pages 477–498. Springer, 2016. 177, 178, 179, 186, 188

W. Thomson. Axiomatic and game-theoretic analysis of bankruptcy and taxation problems: A survey. *Mathematical Social Sciences* 45(3):249–297, 2003. 181

AUTHORS' BIOGRAPHIES

Marianna Belotti, after having completed her studies on Quantitative Finance (Mathematical Engineering) in Politecnico di Milano, joined the Blockchain and Cryptoassets Programme of the French institution Caisse des Dépôts within a CIFRE program (industrial PhD) where she carries out research and development work on projects of strategic importance. Marianna has several publications related to blockchain technology, exploring its potential use-cases and its adaptability and security issues. She is now in charge of the BI operations of the team, as well as the Think Tank activities of LaBChain (a consortium grouping the major actors in the French fintech ecosystem), and she represents the institution within the International Association of Trusted Blockchain Applications (INATBA), where she is co-chair of the Governance working group. Marianna is also an INSEAD alumna since she obtained the Business Foundation Certificate.

Stefano Moretti is a CNRS researcher, and a member (and the deputy director, since 2019) of the LAMSADE laboratory, a joint research center in computer science of the Paris Dauphine University and the CNRS. He graduated in Environmental Science from the University of Genoa in Italy in 1999, and he was awarded a PhD in Applied Mathematics from the same university in 2006. In 2008, he was also awarded a PhD in Game Theory from Tilburg University, The Netherlands. His research activity is organized around the following main lines: the axiomatic analysis of solutions for cooperative games and their applications; the design of efficient algorithms in strategic games arising from combinatorial problems on networks; power indices and related social choice problems.

CHAPTER 8

Tokens and ICOs: A Review of the Economic Literature

Andrea Canidio, *IMT School for Advanced Studies, Lucca*
Vincent Danos, *CNRS, ENS-PSL*
Stefania Marcassa, *CY Cergy Paris University*
Julien Prat, *CNRS, CREST, Ecole Polytechnique*

8.1   INTRODUCTION

Initial coin offerings (ICOs hereafter) are marketed as an alternative to initial public offerings (IPOs) for ventures endeavoring to raise funds. Even though their names are similar, the two processes are quite different. IPOs are usually run by established companies with proven track records in terms of profitability. Moreover, IPOs are associated with the sale of specific equities that are submitted to clear regulations. ICOs are instead organized by startups to raise capital through a token sale directed at a crowd of investors. Often, this token is a cryptocurrency, a digital medium of exchange based on some distributed ledger technology.

The first reported ICO was conducted by Mastercoin in 2013, with the fundraising event running for almost one month. Mastercoin proposed the implementation of a payment system, and raised close to 5,000 bitcoins, valued around $500,000 at the time. Between 2015 and 2017, about 1,000 ICOs followed, raising more than $6 billion from the public, thus vastly exceeding venture capital investments in funding innovative projects related to blockchain technology. For comparison, in 2017, blockchain entrepreneurs raised $5 billion through ICO offerings, as opposed to $876 million through venture capital funding. In 2018, according to ICObench, 2,218 ICOs reached their conclusion and investors could choose, on average, among 482 token sales per day.[1] This initial burst led to tremendous excitement around cryptocurrencies as a new funding model for innovation in the upcoming digital age. But the euphoria proved to be short-lived: outright fraud,

---

[1] Financial estimates as reported by coindesk.com.

rampant speculation, and the resulting volatility in the trading of many cryptocurrencies eventually eroded investor confidence. By the closing months of 2018, the ICO market had recorded a radical downturn.

At the time of writing, the jury is still out on whether ICOs can provide a durable alternative to venture capital. Because ICOs are fundamentally different from other financing arrangements, they cannot be designed or evaluated using off-the-shelf techniques. Instead, ICOs require a new analytical framework, and our contention is that they will not be ranked equally alongside other financing instruments until we have acquired a proper understanding of their specificity. The objective of this survey is to provide readers with an overview of the ongoing efforts toward that goal. Given the burgeoning growth of the field, we make no claim to having been thorough in our coverage of the related literature. In particular, we focus on theoretical insights and refer readers interested in empirical findings to the survey by Li and Mann [2018], and readers interested in the rapid evolution of ICO regulation to the work of Collomb et al. [2018]. We hope that those whose work we have overlooked will understand and forgive.

The chapter is structured as follows. Section 8.2 describes how ICOs are conducted and provides an overview of data on their market history. Section 8.3 discusses the corporate finance issues raised by ICOs and how they should and should not be designed in order to align the interests of sellers and buyers of digital coins. Section 8.4 presents a few models of ICO pricing, while Section 8.5 concludes by discussing the numerous challenges that still have to be tackled.

8.2   DESCRIPTION

Tokens vs. cryptocurrencies

Before describing the process of conducting an ICO, it is important to dispel the common misconception according to which all tokens are substitutes for Bitcoin. Although early blockchain projects, such as Mastercoin, were indeed inspired by Bitcoin, the current wave of ICOs is financing infrastructures and platforms whose scopes go well beyond the realms of payment systems. Accordingly, although cryptocurrencies are indeed tokens, all tokens are not cryptocurrencies. Whereas cryptocurrencies are digital assets used for the sole purpose of making or receiving payments, tokens are virtual representations of an asset or utility within their specific blockchain ecosystem. Take for instance Ether, the token native to the Ethereum platform. Ethers are used to pay the transaction fees required by miners for the execution of the smart contracts recorded on the Ethereum blockchain. Hence Ethers can be consumed to record payments, but they also allow their owners to perform a much broader set of transactions.

The plasticity of tokens has been leveraged by their issuers to raise funds across a vast array of industries and applications. The usual metaphor is that utility tokens are a modern form of vouchers since they entitle their holders to a discount or to the right to exchange them against the services provided by the platform that recognizes them. Finally, native tokens are distinguished from other tokens depending on whether or not they have been issued on the same blockchain as the one that provides the services.[2] Bitcoin and Ether are examples of native tokens. An example of non-native tokens are REP, which allow their owners to access Augur's prediction market, but they have been issued on the Ethereum platform through the execution of a specific custodial contract.[3] To understand how this can be done, we now have to dive into the inner workings of ICOs.

The ICO process

Typically, companies wishing to enter the ICO process will lay down their plans in a public document called a white paper. This document is similar to traditional investment documents for IPOs. It describes the service that the company intends to build, how the associated token relates to that service, how token holders stand to gain, and how the funds obtained will be allocated to various aspects of the project. The service itself can be anything from an online gambling game to be run as a smart contract on an existing blockchain, to an entirely new concept of blockchain. The white paper also describes the parameters of the token sale such as: the starting and closing dates of the sale, the price of the token (often on an increasing schedule), the total amount of tokens minted, and the amount of tokens reserved for the team leading the ICO. Plans for the creation and allocation of tokens during the lifetime of the platform are also laid out. An important additional parameter is how long a token will be locked up before becoming tradable. It is often the case that a substantial amount of tokens are sold privately prior to the public token sale, in which case price and lockup conditions may be different.

The white paper also needs to describe the medium of the sale itself. By and large, ICOs have used specific smart contracts on the Ethereum blockchain to conduct their sales. An investor willing to buy a certain amount of tokens has to obtain a corresponding amount of cryptocurrency and to transfer that amount to the dedicated contract, which will lock up that amount until the closing date. The contract may define a soft and a hard cap. The hard cap fixes a maximum amount of tokens to be sold. The soft one fixes a minimum. If at the closing date the total amount of tokens sold is under the soft cap, the operation is canceled and buyers are refunded. The initial token price is usually written in the cryptocurrency native to the chain that hosts the sale contract.

The fact that investors buy their tokens with a cryptocurrency and that the entire operation can be conducted on the associated chain as a smart contract has interesting consequences. For one thing, the traditional IPO ecosystem of financial advisers and various intermediates is completely

---

[2] There is some disagreement on the terminology as some users prefer to use the term *coins* for assets that are native to their own blockchain, and reserve the term *token* for assets that are created on existing blockchains.

[3] This contract belongs to the often-used ERC-20 class and can be examined at Etherscan [2021].

bypassed. Second, there is an exchange risk as the cryptocurrency is volatile relative to fiat. Because the investment process locks each buyer's payment if the sale is unsuccessful, it may be that the amount refunded has a much lower fiat value than it had at the start. Third, the target audience of the sale is international and the sale therefore addresses a large and diverse community of buyers. Buyers typically go through some KYC (know your customer) process to establish a modicum of identification both for regulatory purposes and for marketing ones. In the case where the service to be built is not hosted on the chain(s) used to collect the investments, there is also a question of whether to trust the proponents of the ICOs. Scams exist, but most of the investment risk lies simply in assessing the value proposition of a project and whether its proponents will execute it as planned. Often promises are either unrealistic and/or not pursued. These concerns have been mitigated by the recent emergence of so-called IEOs (initial exchange offerings). These are bona fide ICOs but the sale process differs. Sales are mediated by a crypto-exchange. Investors are offered a selected set of projects, presumably of higher investment quality, and a promise from the exchange to list the token. The latter may be a critical incentive for short-term traders and indeed the first IEOs have been susbcribed in no time—triggering a rebirth in the world of ICOs.

Market history

Aggregate all-time ICO proceeds reached $22.5 billion by the end of October 2018, with an average ICO size increasing from $4.35 million in 2014 to $25.72 million in 2018. Moreover, ICOs have displaced equity issuance and venture capital as sources of capital for blockchain-based startups.

To give an overview of the evolution of capital raised and geographical distribution of ICOs, we draw on the following two papers: Benedetti and Kostovetsky [2021] and Howell et al. [2018]. Both articles combine data from CoinMarketCap and other websites, as there is currently no industry-standard data source. More precisely, the work in Howell et al. [2018] combines ICOs from TokenData.io and CoinMarketCap.com, while the work in Benedetti and Kostovetsky [2021] collects data from five aggregator websites which, when combined, give the largest number of ICOs, the most available characteristics, and the highest accuracy—icodata.io, ICObench.com, icorating.com, icodrops.com, and icocheck.io—in addition to CoinMarketCap. For the most recent periods, we rely on data from ICObench.com.[4]

The work in Howell et al. [2018] shows that between April 2014 and June 2018, ICOs raised over $18 billion, as shown in Figure 8.1, with a peak in 2017q2 (second quarter of 2017), when at least 15 individual ICOs raised more than $100 million.

After the summer of 2018, the trend changed: ICObench reported a decline in the ICO market, as the number of completed projects in 2019q1 was half that of 2018q4. The overall amount

---

[4] Note, however, that the patterns described here are broadly similar across all the empirical literature and specialized websites that we are aware of.

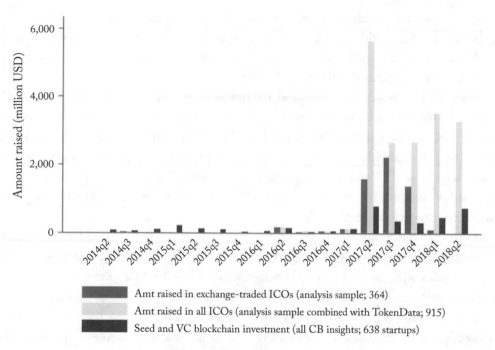

Amt raised in exchange-traded ICOs (analysis sample; 364)
Amt raised in all ICOs (analysis sample combined with TokenData; 915)
Seed and VC blockchain investment (all CB insights; 638 startups)

FIGURE 8.1: Amount raised through ICOs, 2014q2–2018q2. Note: This figure compares the amount raised through ICOs with the amount raised by blockchain-related startups. Data is quarterly from 2014 through the second quarter of 2018. The dark blue bars show total funding in our estimation sample of 453 exchange-traded ICOs (of which amount raised is non-missing for 364). The light blue bars combine our estimation sample with all remaining tokens that had completed ICOs and available amount raised from the TokenData database. Source: Howell et al. [2018].

of funds raised in 2019q1 still totaled about $1 billion, however. Indeed the average funds raised in this same quarter ($8.4 million) were higher than in 2018q4 ($6.8 million), and the percentage of ICOs that raised a positive amount of funds was nearly the same (35 → 33%), even though their number decreased (from 207 to 107).

The change in trend at the end of 2018 concerned ICOs' durations as well. The mean ICO duration was between 37 and 40 days in Benedetti and Kostovetsky [2021] and Howell et al. [2018], and it increased to 97 days in January 2019, according to ICObench. Finally, in Benedetti and Kostovetsky [2021], listing ICOs exhibited a median return of 21%, but a mean return of 246%, from the ICO closing price to the first opening price. Over a longer horizon, the work in Howell et al. [2018] has found that, conditional on being listed on CoinMarketCap, the mean five-month cumulative excess return over Bitcoin is +150%, while the median is −50%.

Even though these numbers cannot be directly compared across sources, as they are sensitive to the details of sample construction, they all reflect a heavily skewed distribution of ICO proceeds, with a small number of highly successful transactions driving much of the aggregate activity. This is not abnormal in the world of startup financing. Venture capital firms make most of their profits from only a few investments. The dotcom bubble, for instance, generated many failed companies, but it also gave rise to giants like eBay and Amazon.

Figure 8.2 [Howell et al. 2018] shows the distribution of ICOs by country as of April 2018. Since these are token sales for decentralized digital platforms, they do not really have a single home country. In fact, the team of entrepreneurs and employees are often from many different countries, and the country of registration or incorporation is usually chosen for legal and tax reasons, which is why noted havens like British Overseas Territories, Singapore, Switzerland, Cyprus, and the Baltic States, especially Estonia, are over-represented in Figure 8.2. The largest English-speaking countries and Russia are also popular locations for ICOs.

In 2019q1, ICObench reported that the United States, Singapore, and the United Kingdom were the countries hosting the largest number of ICOs since 2015, and these three were also the countries with more new ICOs ending during 2018. In 2018, the U.K. overcame Russia in the general ranking, while Germany reached the eighth position, overtaking Canada and the Netherlands—the latter of which exited the top 10. Considering the whole sample of ICOs indicating a precise localization in their white papers, the spatial concentration decreased from 2017 to 2019, but remained very high: ICOs hosted by the top 10 countries accounted for about 76 percent of the capital gathered, and the projects establishing their headquarters in the United States alone amassed almost 30 percent of the funds. In 2017, these values were even higher: 90 percent and 61 percent, respectively. Moreover, in response to risks and potential abuses, China banned cryptocurrencies at the end of 2017 and South Korea has pursued stern regulatory policies.

In this regard, the work in Benedetti and Kostovetsky [2021] shows that ICOs are mostly located in countries with above-average World Bank Rule of Law rankings and high standards of living, two metrics that are highly correlated with each other. They also show that these two measures are positively associated with ICO success, with listed ICOs originating from countries that are 0.2 points higher in their Rule of Law rating and have about $4,000 more in GDP per capita, relative to their entire sample.

8.3   CORPORATE FINANCE OF TOKENS

So far we have discussed the use of tokens as a means to raise funds. But tokens can also be used to generate incentives. For example, as already discussed in previous chapters, blockchain protocols combine cryptographic tools and economic incentives. The main purpose of a protocol's token is precisely to provide those incentives, because it can be used to remunerate the network participants

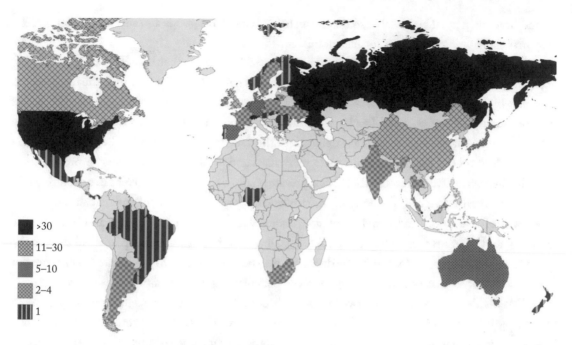

FIGURE 8.2: The geography of ICOs. Note: This figure shows the location of ICO issuers in our sample. Not shown are the Cayman Islands (3 ICOs), Curacao (1), Cyprus (1), Gibraltar (1), the Marshall Islands (1), Saint Kitts (1), and ICOs whose teams are dispersed across >4 countries (28). There are additionally 87 ICOs whose issuer locations are unknown. Source: Howell et al. [2018].

according to rules specified within the protocol itself.[5] Of course, for these tokens to generate incentives, two additional conditions must be satisfied. First, they must have positive value, which is typically achieved by establishing that the token is necessary in order to operate the protocol.[6]

---

[5] Note that this problem is specific to public blockchains. In private blockchains, an organization (either a consortium or a single firm) determines who can participate in the network. Such an organization also provides incentives to the network participants. This is why private blockchains do not need to use tokens.

[6] The simplest way to make a token necessary is to establish it as the internal currency of the protocol (see decentralized marketplaces such as Golem, Sia, Filecoin, or iExec). But the function of a token within a protocol can be extremely complex. For example, MakerDAO is a blockchain-based protocol that allow users to create a collateralized debt obligation (CDO) by locking a given amount of Ethers in a smart contract. The system is designed so that the USD-value of such an obligation is roughly constant. Its associated token is MKR and has several uses. For example, users who want to retrieve their collateral will need to pay a fee established in MKR. Also, if the value of the collateral falls below the nominal value of the CDO, the protocol will automatically liquidate the collateral, and simultaneously sell newly created MKR so as to cover the shortfall. MKR holders are therefore bearing the risk associated with issuing the CDO. Finally, an MKR holder can vote to change some parameters of the protocol (such as the fee required to retrieve a collateral or the threshold below which a collateral is liquidated).

Second, the token must be tradable so that it can be exchanged for goods and services, either directly or by first exchanging the token for fiat money.

By virtue of having positive value and being tradable, a token that is successful in generating incentives within the protocol can also generate incentives *off-protocol*. For example, the team behind a specific blockchain-based protocol may hold a large number of tokens initially. To the extent that the quality of the protocol (scale, reliability, usability, security, expressivity, privacy-related guarantees, governance structures, etc.) determines the value of the protocol's token, then, holding a large stock of tokens provides incentives to work hard and produce a high-quality protocol.

In this section we review the theoretical literature studying the role of tokens in generating off-protocol incentives. We call this body of literature "corporate finance of tokens" because it studies the impact of tokens on traditional corporate finance decisions such as whether and how to raise financing, and what investments to undertake. It is, however, important to keep in mind that, in the context of blockchains and tokens, those decisions may be relevant also to organizations that are not corporations but consist of a loose group of developers working on the same open-source project.

We divide the analysis into two parts, each corresponding to a specific type of token. In the first part, we consider tokens that are associated to blockchain-based protocols. In this case a developer (or a group of developers) releases the protocol as an open-source, free-to-use piece of software that can be utilized by third parties to transact with each other. These transactions are fully disintermediated, in the sense that the developer does not impose fees or prices for using the technology. The developer's source of revenue is the sale of the token that is required to use the protocol. This case covers the largest ICOs to date, as well as tokens representing at least 90% of the total crypto-market.[7]

In the second part, we consider tokens that represent contracts between issuers and the token holders.[8] For example, a token may represent a voucher and therefore give the right to acquire a good or a service from the issuer. Or it may represent a full-fledged security, such as a claim to the issuer's profits or revenues. Importantly, in this case the issuer is typically a company operating a business that may be completely unrelated to blockchain (except for the issuance of tokens). Although quantitatively less important than the first case, tokens belonging to this second category seem poised to play a central role in the future due to their relevance to firms in general, not only to those developing blockchain-based protocols.

Although useful, this classification also has limitations. The main one is that it can be ambiguous whether a specific token falls into one or the other category. The reason is that the

---

[7] At the time of writing, among the top 30 tokens the only ones that are not associated to blockchain-based protocols are Binance Coin (a voucher to access Binance, an exchange) and Tether (a token backed by USD deposits). The value of the remaining 28 tokens is 90% of the total crypto-market (data from http://www.coinmarketcap.com).

[8] Here contract is synonymous with agreement. A separate matter is whether this agreement can be legally enforced. This is, indeed, an area open for additional research. See the conclusions for further discussions.

developer of a blockchain-based protocol could sell the protocol's tokens before the protocol is mature enough to be used, under the expectation (explicit or implicit) that he will complete the work in the future. Establishing whether, in this case, buyer and seller of the token enter into a contractual agreement (and the type of contract) is the core of the debate regarding the regulation of ICOs. It will not, however, be discussed here.[9]

A second limitation is that some of the papers discussed below that are most closely related to one case may be relevant for the other case as well. For example, several authors argued that the way the token is sold can help overcome coordination failures whenever there are network externalities. Indeed, a successful ICO may create the expectation that many people will want to use the protocol, and this expectation may induce even more people to use the protocol (see Li and Mann [2018], discussed below). Network externalities are extremely important for blockchain-based protocols, and hence this insight primarily applies to the first case. But it may also be relevant to the second case if the token issuer is building a platform or any other business with strong network externalities. With these two caveats in mind, we now proceed to discussing the relevant literature.

8.3.1 PROTOCOL ASSOCIATED TOKENS

Articles investigating the role of tokens associated to blockchain-based protocols in generating off-protocol incentives can be roughly divided into three groups: those studying the incentives to develop a blockchain protocol (e.g., Canidio [2018]); those studying the incentives to adopt a blockchain protocol (e.g., Bakos and Halaburda [2019], Cong et al. [2021], Li and Mann [2018], Sockin and Xiong [2020]); and those studying the incentives to maintain a decentralized platform[10] (e.g., Cong et al. [2020]).

To the best of our knowledge, the only work studying the incentives to create a blockchain-based protocol is in Canidio [2018]. In the model, a developer can exert effort and invest funds in the development of a blockchain-based protocol. The developer holds the initial stock of tokens and can choose when to hold an ICO. Following the ICO, in every period there is a frictionless market for tokens where users, investors, and the developer himself can trade tokens. The protocol can be used indefinitely, but the developer has a finite lifespan and hence, at some point, will sell all his tokens and exit the game.

Tokens therefore can be sold to raise funds to invest in the development of the protocol, or they can be sold to earn a profit. The main result is that there is a trade-off between these two uses of tokens. The reason is that in every period in which the market for tokens is open, in equilibrium there is a positive probability that the developer will sell all his tokens and, as a consequence, no

---

[9] See Collomb et al. [2018] for an analysis of the regulatory issues raised by ICOs.

[10] A decentralized platform is the peer-to-peer network generated by the different users of a blockchain-based protocol.

development will occur. Holding an ICO, therefore, allows the developer to raise funds but also introduces the possibility that the developer will dump all his tokens on the market in the future. Interestingly, the equilibrium is inefficient even if the developer has enough of his own funds to invest in the development of the protocol (and hence the only role of tokens is to generate a profit). From the social welfare viewpoint, effort and investment should maximize the present discounted value of the surplus generated by the protocol. The developer instead maximizes the price at which he can sell his tokens, which depends exclusively on usage of the protocol during the period in which he expects to sell his tokens. He therefore ignores completely the fact that the protocol will generate value intertemporally.[11]

Sockin and Xiong [2020], Cong et al. [2021], Bakos and Halaburda [2019], and Li and Mann [2018] have all studied the role of tokens in achieving high-adoption equilibria in decentralized platforms. Because the value of the platform for potential users will depend on the overall level of adoption (network externalities), each person will want to join a decentralized platform only if he or she expects other people to join as well. This scenario may give rise to multiple equilibrium levels of adoption. One distinct feature of a decentralized platform (relative to all other contexts in which there are network externalities) is that all exchanges occurring on the platform must use the protocol-specific (and platform-specific) token. The papers above study how the presence of such a token affects a platform's equilibrium levels of adoption. In Sockin and Xiong [2020], agents first purchase tokens and then trade on a decentralized platform. The authors compare the set of equilibria under full information to the set of equilibria emerging when users do not know other users' demand functions. The main result in Sockin and Xiong [2020] is that the token price and volume are noisy signals for the demand for the platform, and determine what type of adoption equilibrium will emerge. Cong et al. [2021] consider an infinite-horizon, continuous-time model in which agents purchase tokens to trade on a platform. In this environment, holding a token is valuable because it allows a user to transact on the platform and because it may appreciate in value. This second dimension introduces a new, *intertemporal* source of network externalities: holding tokens is more valuable if more users are expected to hold tokens in the future. Relative to a benchmark without tokens, the presence of tokens will therefore "accelerate the adoption of productive platforms or precipitate the abandonment of unproductive platforms."[12]

Bakos and Halaburda [2019] and Li and Mann [2018] have studied how the creator of the platform can sell tokens in a way that induces the high adoption equilibrium. In Bakos and Halaburda

---

[11] Interestingly, depending on the parameter of the model, the developer may be overinvesting or underinvesting in the development of the protocol (relative to the first best).

[12] Cong et al. [2021], page 6. In their model, productivity measures the overall technological quality of the protocol, which evolves stochastically over time.

[2019], a 2-period model is considered in which, in every period, the token issuer will sell tokens and then users will trade on the platform. Period-1 users can purchase tokens only from the issuers, while period-2 users can purchase tokens also from period-1 users. The main result is that the platform owner is always better off *not* introducing tokens, and instead inducing the high adoption equilibrium via negative initial prices (i.e., a subsidy). Tokens will therefore be used only when the platform owner is cash constrained and cannot pay those subsidies. The intuition is that tokens have "equity-like" properties because early adopters benefit from the token appreciation (which depends on the platform's future success), therefore reducing the platform owner's profits. In Li and Mann [2018], the platform owner sells tokens initially in an ICO, and then users use tokens to transact in an infinitely repeated game. The authors show that the sale of tokens at an ICO can be used to induce the high-adoption equilibrium. Because tokens are worthless elsewhere, purchasing them signals the intention to use the platform. A successful ICO sale generates the expectation that those who purchased at the ICO will use the platform and therefore induces participation also of users who did not purchase the tokens at the ICO. Note, however, that this logic may simply shift the coordination problem from the platform adoption phase to the ICO phase, because users will want to purchase tokens only if other users will also purchase them. The work in Li and Mann [2018] shows that holding the ICO over multiple periods and having an increasing price schedule for tokens (as is commonly observed in ICOs) can overcome this second coordination problem.

The work in Cong et al. [2020] builds on Cong et al. [2021] to study the incentives for a platform owner to maintain a decentralized platform. The owner chooses how many tokens to create in each period, and then uses these tokens to pay workers who will increase the value of the platform (by, for example, fixing small bugs or participating on the platform as miners). The choice of how many tokens to create affects the platform owner's payoff in several ways. Creating new tokens increases the monetary supply and therefore decreases the price of tokens. But to the extent that these new tokens are used to pay workers (instead of being consumed by the platform owner), they will increase the value of using the platform and the demand for tokens. The main result is that the platform owner will choose the monetary policy that maximizes the value of the platform. That is, she does not want to issue too many tokens today because this would decrease the price of tokens, the extent to which workers are willing to be paid in tokens, and hence the value of the platform. One can compare this with the work in Canidio [2018], where the developer (also the platform owner) chooses both how many tokens to sell on the market and how much effort to exert in the development of the protocol. That latter model will better fit a situation in which the technology underlying the decentralized platform (that is, the protocol) still needs to be fully developed. In Cong et al. [2020], instead, there is no such effort, and hence the model better fits a situation in which the technology underlying the decentralized platform already exists.

8.3.2   TOKENS AS CONTRACTS

Any company or startup, whether related to blockchain or not, can enter into agreements and issue securities. In some cases these agreements and securities can be traded and exchanged, often in electronic form (think about bonds, shares, vouchers, and coupons). On a superficial level, therefore, blockchain-based tokens are just one of the possible electronic forms that these agreements and securities can take.

In reality, however, tokens have distinct advantages. The first one is that issuing (and then exchanging) agreements and securities in the form of tokens can be done at virtually no cost. This is the reason why ICOs have been so successful and is the subject of much of this chapter. The second advantage is that the space of feasible contracts is enlarged—that is, tokens allow the contracting parties to specify provisions that cannot be specified in a traditional contract. This is achieved thanks to smart contracts, which can be used, for example, to pay token holders automatically if certain conditions are met, or to automatically provide a service to token holders.

There is an emerging literature studying the problem of a firm that can finance its cost of production by selling tokens in an ICO, and that compares ICO financing with traditional equity financing from a venture capitalist (VC) or by debt financing. Within this literature, Catalini and Gans [2018], Chod and Lyandres [2020], and Garratt and van Oordt [2019] assume that tokens represent a pre-sale of output—that is, a firm issues tokens and simultaneously commits to accepting those tokens as a sole means of payment in the future. From the technical viewpoint, these models borrow from the literature on crowdfunding with the difference that, whereas with crowdfunding only users pre-purchase products, in an ICO investors can also purchase tokens with the goal of reselling them (to users) at a profit. Catalini and Gans [2018] consider an entrepreneur who, by paying an initial cost, can start a project of unknown quality. To finance the initial cost, the entrepreneur can sell tokens at an ICO. In this case, he needs to decide the price for tokens and the token-denominated price for its service, as well as the growth in the supply of tokens over time. Catalini and Gans [2018] show that, even assuming that the entrepreneur can commit initially to a monetary policy, traditional equity financing is preferable to ICO financing. The drawbacks to using tokens (relative to equity financing) are even more severe when the entrepreneur cannot commit to a monetary policy. Chod and Lyandres [2020] introduce a risk-averse entrepreneur who can ex-post choose how much to produce. Their central assumption is that VCs are less diversified than ICO investors, because tokens can be bought in (almost) infinitely small amounts. This implies that ICO financing is preferable to equity financing whenever the project is particularly risky (in the sense, for example, of having a very skewed return distribution). Finally, the work in Garratt and van Oordt [2019] assumes that the entrepreneur, in addition to choosing how much to produce, can also invest in a cost-saving technology. The authors' main result is that the ICO can achieve the first-best level of investment in cost saving, which can be achieved neither under equity financing nor under debt financing. This is because the entrepreneur maintains the incentive to reduce the cost of production even if part of the output was pre-sold at the ICO (which is not the case if profits were pledged to third parties).

The article by Malinova and Park [2018] is, to the best of our knowledge, the only one tackling the problem of token design. They consider a model very close to the one in Chod and Lyandres [2020], but instead assume that tokens can represent either a pre-sale of output or a revenue-share agreement. Each of these design choices leads to a different form of inefficiency: revenue sharing leads to underproduction because only a fraction of the revenue (the fraction not pledged to ICO investors) benefits the entrepreneur; pre-sale of output leads to overproduction because the entrepreneur fails to internalize that each unit produced will reduce the equilibrium price of the token (and hence the return to the token holders). Interestingly, when combined together (either in the same token or by issuing two different types of tokens), they can achieve the same outcome as traditional equity financing. In a version of the model with moral hazard, the appropriate design of a token can "beat" equity in the sense of better aligning the incentives of entrepreneurs and investors, and hence leading to higher profits.

8.4   TOKEN PRICING

Having discussed the implications of tokens for corporate governance, we now turn our attention to the other main area of corporate finance, namely the valuation of financial instruments. For reasons that will become clear below, it is important to separate tokens that are only used for transaction purposes from tokens that provide access to a service.

8.4.1   VALUATION OF CRYPTOCURRENCIES

Not surprisingly, given the spectacular growth of Bitcoin price, the bulk of the research on token pricing has endeavored to determine the value of cryptocurrencies. Pinning down the fundamental value of money is a notoriously hard problem with a long tradition. It is well known that multiple equilibria supported by self-sustaining beliefs may arise.[13] Since most people would agree that cryptocurrencies are not likely to replace but simply compete with fiat money, the question can be narrowed down to determining the rate at which agents will be willing to exchange their fiat money for some cryptocurrency. Once the issue has been reframed in such terms, it becomes clear that the indeterminacy of exchange rates originally established by Kareken and Wallace [1981] should apply. Accordingly, the first insight of pricing theory is to provide a theoretical explanation for the extreme volatility of cryptocurrencies. Going beyond this qualitative finding, ongoing research tries

---

[13] See the book by Rocheteau and Nosal [2017] for a comprehensive study of the theory of money, including recent advances in the field of *new monetarist economics*.

to identify the conditions that allowed cryptocurrencies to bootstrap themselves out of the no-trade equilibrium.

Early contributions by Athey et al. [2016] and Bolt and van Oordt [2019] demonstrate that anticipations of future transactional usage may incentivize investors to hoard currencies that are not yet widely accepted. Focusing on the trade-off between central bank money and privately issued digital currencies, Garratt and Wallace [2018] argue that agents trade storage costs against disaster risks. This threat is explicitly modeled in Pagnotta [2020], where the crash risk is determined by miners' investment, giving rise to price-security feedback loops that may amplify or dampen the effect of demand shocks on Bitcoin price. Besides such fundamental factors, price fluctuations are also driven by speculative behavior. Uhlig and Schilling [2018] show that indeterminacy can support a speculative equilibrium where the cryptocurrency is held by agents who anticipate that its value will appreciate in future periods. Biais et al. [2018] devise and estimate an econometric model which separates changes in Bitcoin price that are driven by fundamental news from those that are driven by self-fulfilling expectations. Their estimation indicates that although fundamentals are significant factors, the bulk of return variations on Bitcoin are driven by multiplicative noise.

The articles listed above build on models of dual-currency regimes because they focus on Bitcoin. Hence, they leave aside the ever-growing proliferation of cryptocurrencies described in Section 8.2. One exception is the work by Garratt and Wallace [2018], who also study the cloning process of Bitcoin, casting some doubt on its social value and long-run sustainability. Fernández-Villaverde and Sanches [2019] identify the conditions under which currency competition works. They extend the canonical framework of Lagos and Wright [2005] and find that, for money competition to be consistent with price stability, the cost function associated with the production of private money must satisfy some restrictive requirements. But they also show that these requirements can be relaxed, and privately issued money made consistent with price stability, when an upper bound on the overall supply of each cryptocurrency is enforced by some immutable protocol. The work in Fernández-Villaverde and Sanches [2019] therefore suggests that the commitment power of blockchains was the critical feature that led to the emergence of multiple cryptocurrencies.

8.4.2   VALUATION OF UTILITY TOKENS

Pricing utility tokens is, in principle, more straightforward than pricing cryptocurrencies because utility tokens should trade at a price that reflects the value of the service they give access to. From the standpoint of valuation theory, such rewards are treated as exogenous and utility tokens can be priced in the same way as other assets. That being said, there is no off-the-shelf formula that can be lifted from the asset pricing literature because utility tokens fundamentally differ from standard securities such as debt and equity.

To see why, we now describe a simplified version of the model proposed by Danos et al. [2019]. Consider a platform that issues tokens and commits to exchanging *one unit of service against one token*. There are two markets: (i) a trading market where tokens are bought using fiat money, and (ii) a commodity market where tokens are sold in exchange for the platform's service. The platform has monopoly power on the commodity market, while the price or exchange rate of the token in fiat money, p, is determined on the perfectly competitive and frictionless trading market.

Each period is divided into two subperiods. The trading market opens at the beginning of the period. It allows users to sell and buy tokens at the market price p_t. Then the trading market closes and preference shocks are revealed. The maximum quantity of services desired, or willingness to pay, of each user is randomly drawn from the continuously differentiable distribution function $F(\cdot)$. Using c and d to denote the quantity of services consumed and desired, we assume that the per-period utility function u of users is linear; that is, $u(c; d) = \min(c, d)$.

The timing is crucial and cannot be reversed. Suppose instead that users first observe their willingness to consume and then adjust their token holdings. Since tokens do not bear any interest, users would find it optimal to hold zero tokens at the beginning of the period and the market price would collapse to zero.

Following the same steps as in Danos et al. [2019], one can show that the token price obeys the following law of motion:

$$rp_t = \underbrace{[1 - F(M)](1 - p_{t+1})}_{\text{convenience yield}} + \underbrace{p_{t+1} - p_t}_{\text{capital gain}}, \tag{8.1}$$

where M is the mass of available tokens per user, and r is the user's discount rate. Condition (8.1) decomposes the rate of returns of tokens into two components: a capital gain and a convenience yield. First, token holders benefit from any appreciation in the price of the token as captured by the capital gain. Token holders may also enjoy a convenience yield; with probability $1 - F(M)$ the marginal token is consumed and provides a utility of 1. Since the service is delivered *in exchange* for the token, we also have to take into account the loss of the token and deduct its price from the marginal utility, as captured by the term $-p_{t+1}$ in the convenience yield. This subtraction of the asset price is the fundamental difference between token and share pricing. Whereas shares give their owner the right to a stream of dividends, utility tokens do not yield any benefits until they have been exchanged. Hence their fundamental value is equal to the discounted surplus of the *next trade*.

Condition (8.1) also highlights that, for the price of the token to remain bounded, the convenience yield should eventually be positive. Setting $p_t = p_{t+1}$ in (8.1) yields the token price in steady-state, which we denote by $\hat{p}$,

$$\hat{p} = \frac{1 - F(M)}{r + 1 - F(M)} < 1. \tag{8.2}$$

As expected, the equilibrium price is decreasing in the overall supply of tokens M. More interestingly, the platform services are paid at a price that is *lower* than the marginal utility since $\hat{p} < 1$. This is the main cost of requiring users to pay in tokens instead of fiat money.[14]

To take stock, this basic model has three main insights. First, it identifies the conditions under which tokens are valuable, namely when users need to acquire the services so quickly that they do not have the time to refill their token holdings. Second, it clarifies the cost for the platform of relying on an ICO to raise funds. By issuing tokens, the platform actually commits to selling its product at a discount that compensates users for the opportunity cost of holding tokens instead of interest-bearing securities. Third, it shows how the pricing formula for tokens fundamentally differs from that for other financial instruments because tokens do not generate any dividends until they are exchanged.

In order to capture how token prices evolve over time, Danos et al. [2019] embed their model into a dynamic framework with time-evolving parameters. Their approach endogenizes the velocity of circulation, making it feasible to compute the price at the ICO stage that is consistent with convergence to the long-run equilibria. This model is therefore related to the one in Cong et al. [2021] and Cong et al. [2020], both already discussed in the context of the corporate finance of tokens. The work in Cong et al. [2021] provides a fully microfounded model for the pricing of utility tokens. It is assumed that users need to stake tokens in order to access the platform and let the platform's productivity fluctuate randomly over time. Cong et al. [2021] derive a general pricing formula that depends on the deterministic trend and volatility coefficient of the demand shifters. Their analysis shows that token appreciation may accelerate platform adoption as users internalize network externalities. Cong et al. [2020] describe how the management of tokens through burning and minting can be fine tuned so as to fulfill the incentive constraints of the different stakeholders.

8.5   CONCLUSION

Despite growing interest in the study of tokens and the incentives that they generate, several promising areas of study have so far received little or no attention.

With respect to protocol-associated tokens, the majority of papers have focused on how tokens can help overcome coordination failures in the adoption phase of the protocols. But besides this, all other areas of inquiry seem, as of now, fairly open. For example, competition between different

---

[14] The other loss for the platform is demand rationing since a share $1 - F(M)$ of users would like to consume more than allowed by their token holdings.

protocols is widespread—see the proliferation of cryptocurrencies (Bitcoin, Bitcoin Cash, Bitcoin Gold, Litecoin, Monero, Dash, Zcash, etc.); the proliferation of decentralized computing platforms, each with their own associated token (Ethereum, EOS, Tezos, Cardano, TRON, Ethereum Classic, NEM, etc.); the multiplicity of decentralized payment platforms (Ripple, Stellar, etc.). The intensity of this competition is partly explained by the fact that most of these protocols are open source, which implies that anybody can modify the source code of a given protocol and then create a "fork": a new protocol, with its own development, incompatible with the initial protocol.[15] It is, however, unclear at this point how competition between blockchain-based protocols affects the incentives to develop these protocols, and whether it undermines their long-run sustainability.[16]

A related issue is the likely emergence of multiple token usage and how it could be addressed by the emergence of an intermediary sector. Intermediation could also help foster market liquidity and accompany the growth of functional cross-chain solutions that allow tokens to hop from one chain to the next. Evidently the widespread availability of chain connectors has the potential to transform the dynamics of competition and may favor more collaborative developments, going some way toward the formation of an abstract blockchain as a homogeneous computational medium.

With respect to tokens issued by firms, an important area of inquiry is firms' incentive to honor their initial commitments. In all the papers mentioned above, it is assumed that the firm can indeed commit, possibly because of an existing legal and regulatory framework.[17] But this seems unlikely. The reason is that tokens can be sold to investors worldwide, each investing small amounts. In this environment, no individual investor has the incentive to monitor and then bring to court a firm that, for example, starts accepting means of payments other than the token or declares bankruptcy after paying its management above-market salaries. The cost of coordinating several investors to organize something akin to a class action may be prohibitively high. It is therefore worthwhile exploring to what extent a firm initial commitment is credible even in the absence of legal enforcement.

Yet another issue that is important to study is token design. As already discussed, a security exchanged as a token can look very different from a traditional security. For example, the issuer of a revenue-share agreement could pay its holders weekly or even daily at virtually no cost. Moreover, smart contracts can be used to automate certain parts of the execution of a security contract. Because of this, different security tokens can be pooled together, creating new types of CDOs (collateralized debt obligations), with a smart contract handling the flow of cash to the CDO holders.

---

[15] See Canidio [2018] for a discussion regarding why almost all blockchain-based protocols are open source.

[16] It is worth mentioning that the work in Abadi and Brunnermeier [2018] models competition among blockchains. It is argued that the possibility of creating a fork determines how the blockchain record keepers (i.e., the miners) can be rewarded. Therefore, the authors connect blockchain competition to the design of blockchain-based protocols and on-chain incentives.

[17] The work in Catalini and Gans [2018] also discusses the role of commitment in ICO financing.

Developing the theoretical approaches we have reviewed to address new token designs and the resulting incentives, as well as the pricing of structured tokens, is likely to be one of the most fruitful areas for further research.

BIBLIOGRAPHY

J. Abadi and M. Brunnermeier. Blockchain economics. National Bureau of Economic Research, working paper 25407, 2018. DOI: 10.3386/w25407. 207

S. Athey, I. Parashkevov, V. Sarukkai, and J. Xia. Bitcoin pricing, adoption, and usage: Theory and evidence. Stanford University Graduate School of Business, Research Paper No. 16-42, 2016. 204

Y. Bakos and H. Halaburda. The role of cryptographic tokens and ICOs in fostering platform adoption. CESifo Working Paper Series 7752, CESifo, 2019. 199, 200, 201

H. Benedetti and L. Kostovetsky. Digital tulips? Returns to investors in initial coin offerings. *Journal of Corporate Finance*, Volume 66, 101786, 2021. 194, 195, 196

B. Biais, C. Bisière, M. Bouvard, C. Casamatta, and A. Menkveld. Equilibrium Bitcoin pricing. TSE working paper, 18-973, 2018. 204

W. Bolt and M. van Oordt. On the value of virtual currencies. *Journal of Money, Credit and Banking*, 2019. https://doi.org/10.1111/jmcb.12619. 204

A. Canidio. Financial incentives for open source development: The case of blockchain. MPRA Paper, University Library of Munich, Germany, 2018. https://EconPapers.repec.org/RePEc:pra:mprapa: 85352. 199, 201, 207

C. Catalini and J. S. Gans. Initial coin offerings and the value of crypto tokens. Working Papers Series, 24418, National Bureau of Economic Research, 2018. http://www.nber.org/papers/w24418. 202, 207

J. Chod and E. Lyandres. A theory of ICOs: Diversification, agency, and information asymmetry. *Management Science* (forthcoming), 2020. https://doi.org/10.1287/mnsc.2020.3754. 202, 203

A. Collomb, P. De Filippi, and K. Sok. From IPOs to ICOs: The impact of blockchain technology on financial regulation. SSRN Electronic Journal, 2018. https://dx.doi.org/10.2139/ssrn.3185347. 192, 199

L. W. Cong, Y. Li, and N. Wang. Token-based platform finance. Working paper no. 2019-03-028, SSRN Electronic Journal, 2020. https://dx.doi.org/10.2139/ssrn.3472481. 199, 201, 206

L. W. Cong, Y. Li, and N. Wang. Tokenomics: Dynamic adoption and valuation. *The Review of Financial Studies* 34(3):1105–1155, 2021. 199, 200, 201, 206

V. Danos, S. Marcassa, and J. Prat. Fundamental pricing of utility tokens. THEMA working paper 2019-11, 2019. https://ideas.repec.org/p/ema/worpap/2019-11.html. 205, 206

Etherscan. The Ethereum blockchain explorer. https://etherscan.io/token/0x1985365e9f78359a9b6ad 760e32412f4a445e862. (Accessed 9 July 2021.) 193

J. Fernández-Villaverde and D. Sanches. Can currency competition work? *Journal of Monetary Economics* 106, 1–15, 2019. https://doi.org/10.1016/j.jmoneco.2019.07.003. 204

R. Garratt and M. R. van Oordt. Entrepreneurial incentives and the role of initial coin offerings. SSRN Electronic Journal, 2019. https://dx.doi.org/10.2139/ssrn.3334166. 202

R. Garratt and N. Wallace. Bitcoin 1, Bitcoin 2, . . . : An experiment in privately issued outside monies. *Economic Inquiry* 56(3): 1887–1897, 2018. 204

S. Howell, M. Niessner, and D. Yermack. Initial coin offerings: Financing growth with cryptocurrency token sales. European Corporate Governance Institute (ECGI)—Finance Working Paper No. 564, 2018. 194, 195, 196, 197

J. Kareken and N. Wallace. On the indeterminacy of equilibrium exchange rates. *Quarterly Journal of Economics* 96(2):207–222, 1981. 203

R. Lagos and R. Wright. A unified framework for monetary theory and policy analysis. *Journal of Political Economy 113*, no. 3, 463–484, 2005. DOI: 10.1086/429804. 204

J. Li and W. Mann. Initial coin offerings and platform building. SSRN Electronic Journal, 2018. https://dx.doi.org/10.2139/ssrn.3088726. 192, 199, 200, 201

K. Malinova and A. Park. Tokenomics: When tokens beat equity. SSRN Electronic Journal, 2018. https://dx.doi.org/10.2139/ssrn.3286825. 203

E. Pagnotta. Decentralizing money: Bitcoin prices and blockchain security. *Review of Financial Studies*, ISSN: 0893-9454, 2020. Available at: https://dx.doi.org/10.2139/ssrn.3264448. 204

G. Rocheteau and E. Nosal. *Money, Payments, and Liquidity*. MIT Press, 2017. 203

M. Sockin and W. Xiong. A model of cryptocurrencies. Working Paper No. 26816, National Bureau of Economic Research, 2020. http://www.nber.org/papers/w26816. 199, 200

H. Uhlig and L. Schilling. Some simple Bitcoin economics. NBER Working Paper No. 24483, 2018. http://www.nber.org/papers/w24483. 204

AUTHORS' BIOGRAPHIES

Andrea Canidio is Assistant Professor of Economics at IMT School for Advanced Studies in Lucca (Italy). His research interests include economic development, innovation, political economy, and blockchain. His work on blockchain studies the incentives of blockchain developers, and also competition between blockchain-based platforms. He has contributed to the design of blockchain-based protocols (VeriOSS, the first blockchain-based marketplace for software bug-bounty). He earned his PhD in Economics from Boston University.

Vincent Danos is a computer scientist—currently Researcher at CNRS—with a large body of contributions to the development of logical methods for applied stochastic models, as well as to the foundations of probabilistic programming and inference. He has also made several contributions to the design of models of reversible distributed systems and to game semantics for programming languages. He was awarded an ERC Advanced Fellowship and has experience working with companies in translational research.

Stefania Marcassa is an Associate Professor of Economics at CY Cergy Paris University. She is an applied economist, specializing in labor economics and macroeconomics. She has published her work in international economics journals such as the *BE Journal of Macroeconomics, Explorations in Economic History*, and *IZA Journal of Labor Policy*. She earned her PhD in Economics from the University of Minnesota.

Julien Prat is an economist working on blockchains, mechanism design, contract theory, and macroeconomics. He is a 2004 PhD graduate from the economics department of the European University Institute. He is currently working as a CNRS director of research at CREST, and as an associate professor at the Ecole Polytechnique. Previously, he held assistant professorship positions at the University of Vienna and at the Institute for Economic Analysis (CSIC) in Barcelona. He has published in top international scientific journals such as the *Journal of Political Economy, Journal of the European Economic Association, Journal of Economic Theory*, and *Economic Journal*.

Editors' Biographies

ANTONIO FERNÁNDEZ ANTA

Antonio Fernández Anta is a Research Professor at IMDEA Networks. Previously he was on the faculty of the Universidad Rey Juan Carlos (URJC) and the Universidad Politécnica de Madrid (UPM), where he received an award for his research productivity. He was a postdoc at MIT from 1995 to 1997, and spent sabbatical years at Bell Labs Murray Hill and MIT Media Lab. He was awarded the Premio Nacional de Informática "Aritmel" in 2019 and has been a Mercator Fellow of the SFB MAKI in Germany since 2018. He has more than 25 years of research experience, and more than 200 scientific publications. He was the Chair of the Steering Committee of DISC and has served in the TPC of numerous conferences and workshops. He received his MSc and PhD from the University of SW Louisiana in 1992 and 1994, respectively. He completed his undergraduate studies at the UPM, having received awards at the university and national level for his academic performance. He is a Senior Member of ACM and IEEE.

CHRYSSIS GEORGIOU

Chryssis Georgiou is an Associate Professor in the Department of Computer Science at the University of Cyprus. He holds a PhD (2003) and MSc (2002) in Computer Science and Engineering from the University of Connecticut. His research interests span the theory and practice of fault-tolerant distributed computing with a focus on algorithms and complexity. Recent research topics include the specification and implementation of distributed ledgers, the design and implementation of fault-tolerant and strongly consistent distributed storage systems, and the design and analysis of self-stabilizing distributed systems. He has published more than 100 articles in journals and conference proceedings in his area of study and he has co-authored two books on robust distributed cooperative computing. He has served on several program committees of conferences in distributed and parallel computing and on the steering committees of DISC and ACM PODC (he is currently chairing the steering committee). He served as the General Chair of PODC 2015, the Self-Stabilizing Systems Track Program Committee co-Chair of SSS 2017, the General co-Chair of ApPLIED 2018 and ApPLIED 2019, and the PC co-Chair of NETYS 2020. Since January 2018, he has been on the editorial board of *Information Processing Letters*. Dr. Georgiou's research has been funded by the University of Cyprus, the Cyprus Research and Innovation Foundation, and the European Commission.

MAURICE HERLIHY

Maurice Herlihy has an AB in Mathematics from Harvard University, and a PhD in Computer Science from MIT. He is currently the An Wang Professor of Computer Science at Brown University. He has served on the faculty of Carnegie Mellon University and the staff of DEC Cambridge Research Lab. He is the recipient of the 2003 Dijkstra Prize in Distributed Computing, the 2004 Gödel Prize in theoretical computer science, the 2008 ISCA influential paper award, the 2012 Edsger W. Dijkstra Prize, and the 2013 Wallace McDowell award. He received a 2012 Fulbright Distinguished Chair in the Natural Sciences and Engineering Lecturing Fellowship, and he is a fellow of the ACM, as well as a fellow of the National Academy of Inventors, the National Academy of Engineering, and the National Academy of Arts and Sciences. His research focuses on various aspects of parallel and distributed computing, including linearizability, lock-free and wait-free synchronization, transactional memory, concurrent data structures, and blockchain.

MARIA POTOP-BUTUCARU

Maria Potop-Butucaru has been a full professor at the Sorbonne University since 2012, and she has led the Network and Performance Analysis team in LIP6 laboratory since 2018. She received her BSc in Computer Science in 1996 from University Al. I Cuza, Iasi, Romania, and her MSc in 1997 jointly from University Al. I Cuza and Paris XI University, Orsay, France. She received her PhD in 2000 from Paris XI University, France. She was Associate Professor in University Rennes 1 from 2001 to 2006, then Associate Professor at Sorbonne University (former Pierre and Marie Curie University) from 2006 to 2011. Her research interests are distributed systems resilient to multi-faults and attacks (crash, Byzantine, transient, etc.). She is interested in self* (self-organizing, self-healing, and self-stabilizing) and secure static and dynamic distributed systems (e.g., blockchains, peer-to-peer networks, sensors, and robot networks). She focuses in particular on the conception and proof of dependable distributed algorithms for fundamental distributed computing problems: communication primitives (e.g., broadcast, converge-cast, etc.), self* overlays (various spanning trees, P2P overlays, etc.), coherence and resource allocation problems (storage, mutual exclusion, etc.), consensus or leader election. She is editor of blockchain-related special issues of *Annals of Telecommunications* and *Theoretical Computer Science*. She served as PC member, chair, or general chair for several venues in distributed computing SSS, OPODIS, DISC, PODC, etc. She was deputy director of the Computer Science Master School in Sorbonne University.

Printed in the United States
by Baker & Taylor Publisher Services